CORE COMPETENCIES FOR FEDERAL FACILITIES ASSET MANAGEMENT THROUGH 2020

Transformational Strategies

Committee on Core Competencies for Federal Facilities Asset Management, 2005-2020

Board on Infrastructure and the Constructed Environment

Division on Engineering and Physical Sciences

NATIONAL RESEARCH COUNCIL
OF THE NATIONAL ACADEMIES

THE NATIONAL ACADEMIES PRESS
Washington, D.C.
www.nap.edu

THE NATIONAL ACADEMIES PRESS   500 Fifth Street, N.W.   Washington, DC 20001

NOTICE: The project that is the subject of this report was approved by the Governing Board of the National Research Council, whose members are drawn from the councils of the National Academy of Sciences, the National Academy of Engineering, and the Institute of Medicine. The members of the committee responsible for the report were chosen for their special competences and with regard for appropriate balance.

This study was supported by a series of contracts and grants between the National Academy of Sciences and the sponsor agencies of the Federal Facilities Council. Any opinions, findings, conclusions, or recommendations expressed in this publication are those of the author(s) and do not necessarily reflect the views of the organizations or agencies that provided support for the project.

International Standard Book Number-13:   978-0-309-11400-4
International Standard Book Number-10:   0-309-11400-4

Additional copies of this report are available from the National Academies Press, 500 Fifth Street, N.W., Lockbox 285, Washington, DC 20055; (800) 624-6242 or (202) 334-3313 (in the Washington metropolitan area); Internet, http://www.nap.edu.

Copyright 2008 by the National Academy of Sciences. All rights reserved.

THE NATIONAL ACADEMIES
Advisers to the Nation on Science, Engineering, and Medicine

The **National Academy of Sciences** is a private, nonprofit, self-perpetuating society of distinguished scholars engaged in scientific and engineering research, dedicated to the furtherance of science and technology and to their use for the general welfare. Upon the authority of the charter granted to it by the Congress in 1863, the Academy has a mandate that requires it to advise the federal government on scientific and technical matters. Dr. Ralph J. Cicerone is president of the National Academy of Sciences.

The **National Academy of Engineering** was established in 1964, under the charter of the National Academy of Sciences, as a parallel organization of outstanding engineers. It is autonomous in its administration and in the selection of its members, sharing with the National Academy of Sciences the responsibility for advising the federal government. The National Academy of Engineering also sponsors engineering programs aimed at meeting national needs, encourages education and research, and recognizes the superior achievements of engineers. Dr. Charles M. Vest is president of the National Academy of Engineering.

The **Institute of Medicine** was established in 1970 by the National Academy of Sciences to secure the services of eminent members of appropriate professions in the examination of policy matters pertaining to the health of the public. The Institute acts under the responsibility given to the National Academy of Sciences by its congressional charter to be an adviser to the federal government and, upon its own initiative, to identify issues of medical care, research, and education. Dr. Harvey V. Fineberg is president of the Institute of Medicine.

The **National Research Council** was organized by the National Academy of Sciences in 1916 to associate the broad community of science and technology with the Academy's purposes of furthering knowledge and advising the federal government. Functioning in accordance with general policies determined by the Academy, the Council has become the principal operating agency of both the National Academy of Sciences and the National Academy of Engineering in providing services to the government, the public, and the scientific and engineering communities. The Council is administered jointly by both Academies and the Institute of Medicine. Dr. Ralph J. Cicerone and Dr. Charles M. Vest are chair and vice chair, respectively, of the National Research Council.

www.national-academies.org

COMMITTEE ON CORE COMPETENCIES FOR FEDERAL FACILITIES ASSET MANAGEMENT, 2005-2020

DAVID J. NASH, *Chair,* Dave Nash and Associates, Birmingham, Alabama
WILLIAM W. BADGER, Arizona State University, Tempe
JENNIFER J. COMPAGNI, Compagni Associates, Highlands, New Jersey
DENNIS D. DUNNE, dddunne & associates, Scottsdale, Arizona
MARTIN A. FISCHER, Stanford University, Stanford, California
MICHAEL J. GARVIN, Virginia Tech, Blacksburg
ALEX K. LAM, CoreNet Global, Atlanta, Georgia
KARLENE H. ROBERTS, University of California, Berkeley
DAVID H. ROSENBLOOM, American University, Washington, D.C.
RICHARD L. TUCKER, University of Texas, Austin
JAMES P. WHITTAKER, Facility Engineering Associates, P.C., Fairfax, Virginia
NORBERT W. YOUNG, JR., The McGraw-Hill Companies, Inc., New York, New York

Staff

LYNDA STANLEY, Director
KEVIN LEWIS, Senior Program Officer
DANA CAINES, Financial Associate
JENNIFER BUTLER, Financial Assistant

BOARD ON INFRASTRUCTURE AND THE CONSTRUCTED ENVIRONMENT

HENRY J. HATCH, *Chair*, U.S. Army Corps of Engineers (retired), Oakton, Virginia
MASSOUD AMIN, University of Minnesota, Minneapolis
REGINALD DesROCHES, Georgia Institute of Technology, Atlanta
DENNIS D. DUNNE, dddunne & associates, Scottsdale, Arizona
PAUL FISETTE, University of Massachusetts, Amherst
LUCIA GARSYS, Hillsborough County, Florida
THEODORE C. KENNEDY, BE&K, Inc., Birmingham, Alabama
SUE McNEIL, University of Delaware, Wilmington
DEREK PARKER, Anshen+Allen Architects, San Francisco, California
WILLIAM WALLACE, Rensselaer Polytechnic Institute, Troy, New York
CRAIG ZIMRING, Georgia Institute of Technology, Atlanta

Staff

LYNDA STANLEY, Director
KEVIN LEWIS, Senior Program Officer
DANA CAINES, Financial Associate
JENNIFER BUTLER, Financial Assistant

Preface

The U.S. government is faced with the ever-increasing challenge of managing the facilities and infrastructure required to support the accomplishment of its missions. Shrinking budgets, the increasing cost of operations and maintenance, changing government priorities, a workforce quickly approaching retirement without equally qualified replacements, and the increasing world focus on sustainable development demand new approaches to federal facilities management.

The Federal Facilities Council of the National Research Council has sponsored several studies looking for insight into possible, plausible solutions to cope with the varied dimensions of the facilities management challenge. Three previous studies focused on, respectively, stewardship of the federal physical infrastructure, outsourcing of management functions, and best practices for investment. Now this fourth study focuses on the people and the skills the federal workforce will need to manage facilities in the next decade and beyond.

This committee's task was to help ensure effective federal facilities management in the next 15 years. Additionally, it was asked to identify effective strategies and processes to ensure that the required core competencies—the essential areas of expertise and the skills base—for facilities management are developed and sustained.

The committee first examined the current situation and the various policies and trends affecting facilities management in the federal government. It reviewed government-wide initiatives and private sector efforts to identify the best ways to proceed in transforming federal facilities management. The committee also reviewed the forces affecting the recruiting, retention, and training of facilities professionals in the public sector. Internal forces include budget reductions, changing government paradigms, the impending loss of many more senior staff through retirements, and the difficulty of attracting new professionals to the federal sector. External forces include the increasing worldwide focus on sustainability and life-cycle facilities management and the impacts of technology.

Once it understood the breadth of the challenge in facilities management, the committee embarked on a review of what various national and international entities and academics are saying about the evolution of facilities management as a profession. This review emphasized the competencies—knowledge, skills, and abilities—required of both organizations and individuals. Many of the skills identified as critical for facilities asset management are significantly different from those identified as critical just 10 years ago. The committee concluded that developing the skills to deliver a life-cycle facilities management approach is indeed a complex assignment and must be a priority for senior government leaders and their organizations.

Based on its understanding of the current environment, the forces affecting the future of facilities management, and its review of what experts are saying and doing, the committee developed recommendations for moving forward. These recommendations address the assessment of core competencies, barriers to recruiting people with the requisite skills, strategies to develop the workforce of the future, creation of an operating environment that supports professional development, and the sustainment of core competencies. Finally, the committee recommended ways to measure progress toward developing and sustaining the skilled workforce that is needed.

In summation, the committee has identified the core competencies it believes are needed for effective federal facilities management and has made recommendations for consideration by decision makers at all levels of government. It trusts that this study has met the objectives set forth by the Federal Facilities Council and, as such, will further the debate and transformation of facilities asset management within the federal government.

David J. Nash, *Chair*
Committee on Core Competencies for
Federal Facilities Asset Management,
2005-2020

Acknowledgments

This report has been reviewed in draft form by individuals chosen for their diverse perspectives and technical expertise, in accordance with procedures approved by the National Research Council's (NRC's) Report Review Committee. The purpose of this independent review is to provide candid and critical comments that will assist the institution in making its published report as sound as possible and to ensure that the report meets institutional standards for objectivity, evidence, and responsiveness to the study charge. The review comments and draft manuscript remain confidential to protect the integrity of the deliberative process. We wish to thank the following individuals for their review of this report:

Doug Christensen, Brigham Young University,
Steve Condrey, University of Georgia,
G. Edward DeSeve, University of Maryland,
Paul H. Gilbert, Parsons, Brinckerhoff, Quade, and Douglas, Inc.,
Vald Heiberg, III, Heiberg Associates,
Sue McNeil, University of Delaware,
David Skiven, General Motors Worldwide Facilities Group,
E. Sarah Slaughter, Massachusetts Institute of Technology, and
Hans A. Van Winkle, Hill International, Inc.

Although the reviewers listed above have provided many constructive comments and suggestions, they were not asked to endorse the conclusions or recommendations, nor did they see the final draft of the report before its release. The review of this report was overseen by Richard N. Wright, National Institute of Standards and Technology (retired). Appointed by the NRC, he was responsible for making certain that an independent examination of this report was carried out in accordance with institutional procedures and that all review comments were carefully considered. Responsibility for the final content of this report rests entirely with the authoring committee and the institution.

The committee also acknowledges and appreciates the contribution of the members of the NRC Board on Infrastructure and the Constructed Environment (BICE). BICE was established in 1946 as the Building Research Advisory Board. It brings together experts from a wide range of scientific, engineering, and social science disciplines to discuss potential studies of interest, develop and frame study tasks, ensure proper project planning, suggest possible reviewers for reports produced by fully independent ad hoc study committees, and convene meetings to examine strategic issues. Dennis Dunne of BICE was a member of the Committee on Core Competencies for Federal Facilities Asset Management, 2005-2020, and Sue McNeil was a report reviewer. None of the other board members listed on page vi were asked to endorse the committee's conclusions or recommendations or to review the final draft of the report before its release.

Contents

SUMMARY 1

1 CONTEXT 13
   The Evolution of Facilities Management, 14
   Facilities Asset Management, 16
   Government-wide Initiatives for Management Reform, 18
   Defining the Problem, 20
   Statement of Task, 21
   The Committee's Approach, 21
   Previous Studies of the National Research Council, 22
   Organization of the Report, 23
   References, 23

2 FORCES AFFECTING THE FEDERAL GOVERNMENT: 25
   IMPLICATIONS FOR FACILITIES ASSET MANAGEMENT
   IN 2020
   Geopolitical and Socioeconomic Trends, 25
   Changing Government Paradigms, 26
   Budgetary Pressures, 28
   Advances in Technologies, 30
   Sustainable Development, 31
   Aging Federal Workforce, 32
   Attracting a New Generation of Workers to the Federal Government, 33
   A New Paradigm Is Essential, 34
   References, 35

3 CORE COMPETENCIES FOR FEDERAL FACILITIES   37
   ASSET MANAGEMENT
   Facilities Management Competencies Literature, 38
   Facilities Management Competencies Identified by Professional
      Associations, 40
   Review of Educational and Professional Development Programs, 42
   Ideas of Federal Agencies on Competencies Required for Facilities
      Management, 42
   Leadership Skills, 45
   Summary, 51
   Required Core Competencies, 51
   Identifying Core Competencies for Facilities Asset Management
      Divisions, 53
   References, 55

4 A COMPREHENSIVE STRATEGY FOR WORKFORCE   57
   DEVELOPMENT
   Creating an Environment for Promoting and Sustaining Core
      Competencies, 57
   Elements of a Comprehensive Workforce Development Strategy, 59
   Strategies for Professional Development, 62
   Performance Measurement, 67
   References, 71

5 CORE COMPETENCIES FOR FEDERAL FACILITIES ASSET   72
   MANAGEMENT: FINDINGS AND RECOMMENDATIONS
   Findings, 72
   Required Core Competencies, 77
   Recommendations, 78
   References, 79

APPENDIXES

A Biographies of Committee Members   83
B Committee Interviews and Briefings   88
C Executive Summary from *Stewardship of Federal Facilities*   89
D Executive Summary from *Outsourcing Management Functions*   99
E Executive Summary from *Investments in Federal Facilities*   110

Summary

Some momentous events in the past several years have brought home the essential role of facilities and infrastructure in supporting the daily operations of businesses and governments and the quality of life for Americans: the attacks on the World Trade Center and the Pentagon in September 2001; the anthrax crisis of November 2001; the blackout in the Northeast in August 2003; and the aftermath of Hurricane Katrina in 2005. In each event the failure of facilities and infrastructure impacted human life, health, welfare, and safety and the provision of essential operations and services. Facilities have a similarly large impact on the environment, accounting for 40 percent of all energy use in the United States and 40 percent of all atmospheric emissions, including the greenhouse gases that have been linked to global climate change. As the 21st century progresses, buildings and infrastructure that are efficient, reliable, cost effective, and sustainable will become even more important.

With a total portfolio of more than 500,000 facilities (NRC, 2004), the U.S. federal government has a significant role to play in reducing their environmental impacts. This portfolio is aging and deteriorating, and it is not aligned with current federal missions: Some facilities are no longer needed, some are poorly located, and others are obsolete. The condition of some federal facilities places the health, safety, and welfare of people at risk, hinders the accomplishment of various missions, and incurs a significant, long-term financial burden (GAO, 2003). Funding for the operation and maintenance of federal facilities, a long-standing issue, will become even tighter if the Government Accountability Office's (GAO's) projected decreases in federal discretionary funds through 2040 come to pass (GAO, 2006).

Until the 1990s, most large organizations that owned facilities managed them

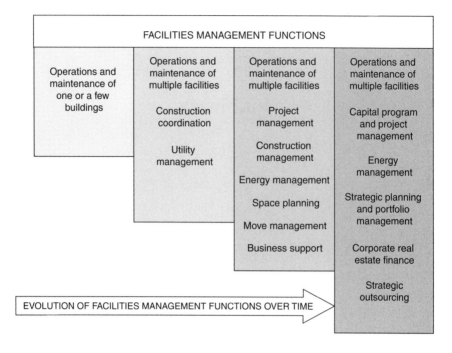

FIGURE S.1 Evolution of facilities management functions. SOURCE: Adapted from APPA (2002).

through an in-house building or engineering division, which typically focused on the tactical issues that arose in the day-to-day operation of individual buildings. As new functions were assigned to the in-house facilities division—strategic planning, construction coordination, utility management, space planning, and project management, among others—the essential areas of expertise and the skills base required to discharge these responsibilities broadened to include financial management and business-related skills (Figure S.1).

The recognition of facility costs, their impact on business operations, workforce health and safety, and the environment, in combination with technological, geopolitical, and socioeconomic trends, is driving a paradigm shift in how public and private organizations manage facilities and in the skills and capabilities required of the people who manage them. This paradigm shift is referred to as "facilities asset management." In this report, facilities asset management is defined as a systematic process for maintaining, upgrading, and operating physical assets cost effectively. It combines engineering principles with sound business practices and economic theory and provides tools to facilitate a more organized, logical approach to decision making (NRC, 2004).

The goal of facilities asset management is to give an organization the work environments it needs to achieve its missions by optimizing available resources. This can best be done by integrating people, places, processes, and technologies to optimize the value of facilities throughout their life cycles—that is, from planning, design, and construction through operations and maintenance, renewal, and ultimately disposal. Effective facilities asset management requires a professional workforce with both hard and soft skills and capabilities in business—including technical disciplines, communications, negotiation, critical thinking, and leadership—and in enterprise knowledge, defined as a profound understanding of an organization's missions, culture, clients, and relationships.

Within the federal government, facilities investment and management decisions involve multiple stakeholders (Congress, the executive branch, the public), decision makers (the President, Congress, the Office of Management and Budget, senior executives of departments and agencies), and more than 30 facilities asset management divisions.[1] The President and Congress are responsible for providing leadership and vision, setting policy, enacting legislation, establishing regulations, and authorizing and appropriating public funds. Civil service employees and political appointees within the various federal organizations are responsible for administering programs, establishing and executing processes, analyzing their results, recommending initiatives, enforcing regulations, and expending public funds efficiently, effectively, and legally (NRC, 2004). In this complex operating environment, facilities asset management divisions must serve as "connected integrators" of diverse stakeholders, functions, and services.

Spurred by internal and external forces, federal organizations[2] are moving toward a facilities asset management approach and transforming the processes used to acquire, manage, operate, invest in, and evaluate their facilities. Their objectives are to strategically align their facilities portfolios with their current and future missions; to ensure the continuity of essential government operations in an emergency; and to do so cost-effectively, efficiently, reliably, and sustainably. However, if a facilities asset management approach is to be fully realized in the federal government, a concurrent transformation must occur in the core competencies—the essential areas of expertise and the skills base required to achieve an organization's missions—of facilities asset management divisions.

Transforming the core competencies of facilities asset management divisions will be challenging. Within the entire executive branch, 42 percent of the Senior Executive Service, which includes most managerial and policy positions, is projected to retire by 2010. A potentially enduring result of these retirements is the

---

[1] A facilities asset management division is defined as the operational unit within a federal organization whose primary responsibility is the management of that organization's portfolio of facilities. Examples of facilities asset management divisions are the Office of Overseas Buildings Operations in the Department of State, the Engineering and Real Property Division in NASA, and the Public Buildings Service in the General Services Administration.

[2] Cabinet-level departments (e.g., the Departments of Defense, State, and Veterans Affairs) and independent agencies (e.g., the General Services Administration and NASA).

loss of the institutional knowledge acquired over decades regarding the design, construction, and operation of specialized facilities, particularly for military and complex civil works projects. Attracting a new generation of workers into federal service—a generation that is savvy about new technology and that was born in the Information Age—will not be easy. This generation will be accustomed to connectivity, mobility, and information availability. Surveys show that this generation does not consider the federal government an "employer of choice" (PPS, 2006). In addition, the federal hiring process is cumbersome, confusing, and slow, and many who do apply for positions drop out of the process to take other jobs. Perhaps most important, the job descriptions for facilities asset managers are based on an outdated paradigm and are not written to attract the professional, highly skilled workforce required to manage federal facilities effectively in the future.

Nonetheless, the current situation also presents a once-in-a-generation opportunity to transform the government's approach to facilities management. As workers retire, federal organizations can redefine their facilities asset management core competencies, rewrite job descriptions, provide support and training to enhance the capabilities of current staff, and hire new staff to fill skills gaps. By taking these steps, federal organizations can develop and sustain a workforce that embodies the knowledge, skills, and abilities required to effectively manage federal facilities through 2020 and beyond.

STATEMENT OF TASK

At the request of the Federal Facilities Council,[3] the National Research Council (NRC) appointed a multidisciplinary committee of experts from the public and private sectors and academia to undertake a study to help ensure effective federal facilities asset management (inclusive of property development and financial and operational functions) in the next 15 years. Critical elements of the study include identifying the organizational capabilities and individual skills required for effective facilities asset management. Other equally critical elements include identifying strategies and processes to ensure development of the required core competencies over a sustained period of time and performance indicators for measuring progress in workforce development.

REQUIRED CORE COMPETENCIES

Based on a literature review, briefings, interviews, current geopolitical and

---

[3]The Federal Facilities Council (FFC) is a cooperative association of more than 20 federal agencies with responsibilities for large portfolios of facilities. The FFC's mission is to identify and advance technologies, processes, and management practices that improve the performance of federal facilities over their life cycles, from planning to disposal. The FFC operates under the aegis of the National Research Council, the operating arm of the National Academy of Sciences, the National Academy of Engineering, and the Institute of Medicine.

socioeconomic trends, and the experience and knowledge of its members, the committee concluded that three areas of expertise are essential for federal facilities asset management divisions through 2020 and beyond, as follows:

- *Integrating* people, processes, places, and technologies by using a life-cycle approach to facilities asset management;
- *Aligning* the facilities portfolio with the organization's missions and available resources; and
- *Innovating* across traditional functional lines and processes to address changing requirements and opportunities.

The required skills base includes a balance of technical, business, and behavioral capabilities and enterprise knowledge. Technical capabilities in fields such as engineering and architecture are essential for life-cycle facilities management. Technical capabilities include knowledge of design and construction; facilities-related systems and their operations and maintenance; acquisition and project management processes; regulations and procedures; information technology and building technology; and analytical skills. Business capabilities include strategic planning and resource management to support an organization's missions. Behavioral capabilities involve the leadership, communication, negotiation, and change management skills required to integrate functions, people, and processes across traditional lines and the capacity to innovate within a dynamic operating environment. Enterprise knowledge includes an understanding of the facilities portfolio and how to align it with the organization's missions; of the organization's culture, policy framework, and financial constraints; of agency inter- and intradependencies; and of the workforce's capabilities and skills. Facilities asset managers[4] will require technical skills, enterprise knowledge, behaviors, and other personal characteristics that allow them to work in a team-oriented environment and to support achievement of a life-cycle approach for facilities asset management. Together, the three essential areas of expertise and the skills base constitute the core competencies for federal facilities asset management divisions through 2020 and beyond.

The committee's findings and its recommendations for developing and sustaining core competencies follow.

FINDINGS

Finding 1: Previous NRC reports on the management of federal facilities recommended that federal organizations approach facilities asset man-

---

[4]An individual in a facilities asset management division whose primary responsibilities involve some aspect of managing the organization's portfolio of facilities. Such individuals include facilities program managers, planners, architects, engineers, and project, operations, or maintenance managers.

agement with the mindset of an owner; align their facilities portfolios to their organization's missions through strategic decision making; take a life-cycle management approach; and measure performance to continuously improve facilities management.

Finding 2: Geopolitical and socioeconomic trends, rapid advances in technology, reliance on outside contractors, budget pressures, heightened focus on sustainable development, and government-wide reforms require a paradigm shift in both federal facilities management and the core competencies of facilities asset management divisions.

Finding 3: Significant reductions in the federal workforce through across-the-board cuts and hiring freezes have resulted in a workforce whose skills are not aligned with new technologies or business practices. These deficiencies will be exacerbated within the next few years as the most experienced facilities managers—the baby boom generation—retires.

Finding 4: Efforts to recruit recent graduates and experienced professionals to federal service are hampered by the poor image of the federal government as an employer, cumbersome hiring processes, and outdated job descriptions. The GS-1600 series, which is the basis for hiring and compensating facilities managers, is based on an old paradigm and does not reflect new realities or required core competencies.

Finding 5: Of all the stakeholders involved in funding, programming, designing, constructing, operating, and maintaining federal facilities, facilities asset management divisions are the only ones involved in all phases. To effectively support their organization's missions, facilities asset managers must integrate people, processes, technologies, services, and knowledge.

Finding 6: The federal operating environment is dynamic and requires that facilities asset management divisions be able to innovate to address changing functional requirements and to take advantage of opportunities for improvement as they arise. Leadership skills—the ability to influence beyond one's authority—are essential.

Finding 7: To ensure that core competencies are established and sustained, federal organizations need a comprehensive workforce development strategy. They will also need to provide a long-term commitment to and investment in the professional development of facilities asset managers.

Finding 8: Where information gaps exist within a facilities asset management division, facilities asset managers will need to find ways to import research-based and experience-based knowledge.

Finding 9: A system for measuring progress in developing and sustaining core competencies is a critical element of a comprehensive workforce development strategy.

RECOMMENDATIONS

Recommendation 1: To effectively manage federal facilities portfolios through 2020 and beyond, federal organizations and their facilities asset management divisions should operate within the overall framework depicted in Figure S.2.

The recommended framework advises federal organizations to

- Adopt the mindset of an owner of facilities;
- Adopt behaviors that integrate facilities-related decisions into strategic planning processes to support the organization's overall missions;

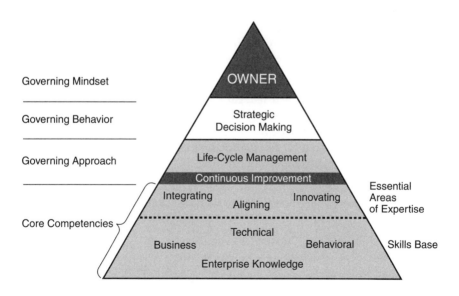

FIGURE S.2 Recommended framework for effective federal facilities asset management.

- Use a life-cycle management approach to operate efficiently, reliably, and cost effectively; and
- Measure performance to support continuous improvement of facilities asset management processes.

To fully implement a facilities asset management approach, federal facilities asset management divisions will require a new set of core competencies. The core competencies comprise essential areas of expertise and a skills base. The three essential areas of expertise are *integrating* people, processes, places, and technologies by using a life-cycle approach to facilities asset management; *aligning* the facilities portfolio with the organization's missions and available resources; and *innovating* across traditional functional lines and processes to address changing requirements and opportunities. The skills base includes a balance of technical, business, and behavioral capabilities and enterprise knowledge.

Recommendation 2: To develop its core competencies, each federal facilities asset management division should first identify the functions and skills it will need to perform or oversee in support of its organization's missions. Although similarities will emerge, the unique aspects of each organization will become apparent only after a careful analysis of current and future functional requirements.

Examples of the types of functions and skills that might be required are shown in Table S.1.

Each facility asset management division must first identify the functions it will need to perform in support of its organization's missions. Once the functions are identified, it should be determined if the division is structured so it can effectively implement a life-cycle management approach or if some reorganization will be required.

Recommendation 3: Each federal facilities asset management division should also conduct an analysis that compares its current skills base to the skills base required for facilities asset management in 2020. The analysis should account for planned or potential changes in missions, requirements, and technologies through 2020 and should identify actions needed to close any skills gaps.

Facilities asset management divisions will need to ask and answer the following questions:

1. What skills are possessed by the facilities asset managers who currently work in the facilities asset management division? Are these skills accessible/available in sufficient quantity for effective facilities asset

TABLE S.1 Examples of Functions and Skills That Might Be Required to Support an Organization's Missions

Technical	Business	Behavioral	Enterprise Knowledge
Operations and maintenance	Strategic planning	Leadership	Mission
Planning and design	Asset management	Teamwork/team building	Vision
Building systems	Finance and accounting	Interpersonal relationships	Strategic direction
Project management	Contract monitoring	Mentoring/coaching	Values
Construction	Procurement	Negotiating	Culture/trust
Code compliance	Real estate	Critical thinking	Systems
Cost estimating	Acquisition and leasing	Communication	Processes
Space planning	Business lexicon	Change management	Resource allocation
Environmental health and safety	Risk management	Quality and innovation	
Energy management	Contingency planning	Future issues/trending	
FM technology	Ethics/law	Performance measurement	
Sustainability	Marketing	Benchmarking	
Commissioning	Human resources		
Security	Professional development		
Life-cycle analyses	Organizational planning		

management? Which skills are found in people who are eligible to retire within 5 years?
2. Is there a gap between the current skills base and the skills base required for effective facilities asset management in 2020?
3. What steps should be taken to close the skills gap? Which skills and capabilities are best acquired through new hires? Which by contracting out? Which by training current staff? Which through a recognized career path for facilities asset managers?

Recommendation 4: Federal organizations should develop a comprehensive, long-term strategy to acquire, develop, and sustain a workforce with the required core competencies for facilities asset management. Senior executives should show their commitment and provide the resources necessary for individuals to develop and refine their skills, including leadership skills, through a continuum of experience and opportunities.

The development and sustainment of a workforce that can fulfill the core competencies required for effective facilities asset management require leadership at all levels of an organization, the sustained investment of resources, and a system to measure progress toward developing workforce skills and capabilities. A comprehensive strategy for workforce development entails recruiting, training, and retaining outstanding individuals. A professional development strategy must go beyond training seminars. It should provide opportunities for education through degree programs or online courses, mentoring, professional certification, participation in professional societies, and knowledge development through research.

Recommendation 5: To overcome barriers to recruiting and hiring individuals with required skills and capabilities, the directors of federal facilities asset management divisions and senior real property officers should collaborate with the Chief Human Capital Officers Council and the Office of Personnel Management to revise the GS-1600 job classification series.

Attracting recent graduates or experienced professionals with the skills and capabilities required to fulfill core competencies for facilities asset management will require significant recruiting and marketing efforts to overcome the negative image of the government as an employer. Recruiting efforts should highlight the full array of public service benefits as well as opportunities that might not be available in private sector firms.

In addition, the job descriptions used to advertise openings and to evaluate potential hires should be updated to reflect the new core competencies. It will probably also be necessary to revise classification and compensation levels to attract high-quality candidates. Such an effort will require cooperation and coordination among facilities asset management divisions, senior real property officers, the Chief Human Capital Officers Council, and the Office of Personnel Management.

Recommendation 6: Federal facilities asset managers should seek to expand the knowledge base related to facilities asset management and use the results to improve decision making and achieve the desired outcomes. Knowledge can be transferred through involvement in professional societies, certification programs, and research using in-house or outside expertise.

Because the discipline of facilities asset management is evolving, there is less conventional wisdom, as well as fewer acknowledged best practices, available than for more established disciplines. Knowledge about facilities asset management can be collected both internally and externally through a variety of activities, including education and training, membership in professional societies, and participation in research.

Recommendation 7: Federal organizations should use a Balanced Scorecard approach for measuring progress in developing and sustaining core competencies for facilities asset management through 2020 and beyond.

Because the Balanced Scorecard (BSC) is a well-established performance measurement system that federal organizations are already applying to some aspects of facilities asset management, the committee recommends its continued use for measuring progress in developing core competencies. The BSC concept has evolved over time, but the four categories of performance have remained the same: financial outcomes, internal business processes, customer relationships, and learning and growth. The learning and growth category focuses on an organization's workforce and its capacity to enable the achievement of the organization's missions. Learning and growth objectives might include enhancing employee skills and capabilities, increasing employee satisfaction, developing an environment for learning, and enhancing the organization's ability to recruit and retain employees with required skills. The indicators to measure progress in achieving these objectives would appear to be a natural outgrowth of the skills gap analysis.

REFERENCES

APPA (Association of Higher Education Facilities Officers). 2002. Development of the Facility Management Profession. Alexandria, Va.: APPA.

GAO (Government Accountability Office). 2003. High Risk Series: Federal Real Property. Washington, D.C.: GAO.

GAO. 2006. The Nation's Long-Term Fiscal Outlook. Washington, D.C.: GAO.

NRC (National Research Council). 2004. Investments in Federal Facilities: Asset Management Strategies for the 21st Century. Washington, D.C.: The National Academies Press.

PPS (Partnership for Public Service). 2006. Back to School: Rethinking Federal Recruiting on College Campuses. Washington, D.C.: Partnership for Public Service.

1

Context

The role of facilities and infrastructure in supporting the day-to-day operations of business and government and improving our quality of life is made apparent by a number of momentous events: the attacks on the World Trade Center and the Pentagon in September 2001; the anthrax crisis of November 2001; the blackout in the Northeast in August 2003; and the aftermath of Hurricane Katrina in 2005. When facilities failed and infrastructure collapsed, the impact on human health, welfare, and safety and on the provision of essential operations and services was evident. Facilities have a large impact on energy usage and the environment as well, accounting for 40 percent of all U.S. energy use and 40 percent of all emissions to the atmosphere, including the greenhouse gases linked to global climate change. As the 21st century progresses, buildings and infrastructure that are efficient, reliable, cost effective, and sustainable will become even more important.

The U.S. federal government is the world's largest single owner of facilities, with a worldwide portfolio of more than 500,000 buildings, structures, and associated infrastructure (NRC, 2004). Comprising billions of square feet of space, these facilities were acquired over more than 200 years to support federal missions and programs ranging from national defense and foreign policy to scientific and medical research, space exploration, cultural arts and history, and recreation. These public assets are valued in the hundreds of billions of dollars. Upwards of $40 billion is spent each year to acquire new facilities and renovate, operate, and maintain existing ones (NRC, 2004).

Nonetheless, many facilities are deteriorating owing to aging and to inadequate maintenance and repair. This is significant because their poor condition hinders performance and the achievement of federal missions. Further, as federal policies, programs, missions, and processes evolve in response to the end of the

Cold War, the 9/11 attacks, and other events, it is clear that facilities portfolios are no longer well aligned with current missions. Federal organizations (departments and agencies) have too many facilities, facilities in the wrong locations, and insufficient resources to operate and maintain them. The Base Realignment and Closure (BRAC) process of the Department of Defense is the most visible and dramatic action taken to align facilities portfolios with missions. However, other federal organizations, including the Department of Veterans Affairs and the Department of State, are undertaking significant, if less visible, actions to realign their portfolios to suit current missions and geopolitical conditions.

Throughout the federal government, the ways in which federal organizations acquire, manage, invest in, and evaluate facilities are being transformed. This transformation requires a concurrent transformation in the core competencies—the essential areas of expertise and the skills base required of the workforce responsible for facilities management.

THE EVOLUTION OF FACILITIES MANAGEMENT

The profession of facilities management is changing rapidly and will continue to evolve through 2020 and beyond. Much of this change has occurred in parallel with the information technology revolution and with the increased expectations of facility owners and users for building performance and cost effectiveness.

Until the 1990s, most of the large public and private organizations that owned facilities managed them through an in-house building or engineering division, often led by a professional engineer. This in-house division typically focused on the tactical issues that arose in the day-to-day operation of individual buildings. An essential area of expertise was building systems operations—mechanical, electrical, plumbing, heating, ventilation, air conditioning, and safety—which required a workforce with technical skills. As new functions were assigned to the in-house facilities division—strategic planning, construction coordination, utility management, space planning, and project management, among others—the essential areas of expertise and the skills base required to discharge these responsibilities broadened to include financial management and business-related skills (Figure 1.1).

In the 1990s, private sector corporations began to "reengineer" their business processes and organizational structures to increase their profitability by becoming more competitive and significantly improving critical areas of performance such as quality, cost, delivery time, and customer service. Three underlying premises of reengineering were that the essential areas of expertise or core competencies of an organization should be limited to a few activities that are central to its current focus and future success; that because managerial time and resources are limited they should be focused on the organization's core competencies; and that services or functions required by the organization that are not core competencies can be outsourced to organizations that specialize in those services. Ideally, by outsourc-

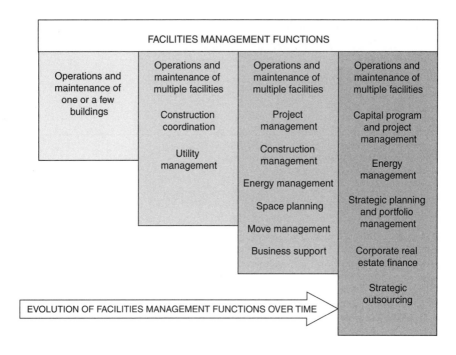

FIGURE 1.1 Evolution of facilities management functions. SOURCE: Adapted from APPA (2002).

ing noncore functions an organization receives the best value or best performance of the resources expended (NRC, 2000).

One result of process reengineering has been closer attention to the function of the facilities and to their costs. As shown in Table 1.1, facilities typically account for almost 25 percent of a corporation's assets and are the second or third highest operating cost after people (salaries and benefits) or after people and information technologies (NRC, 2004).

Recognition of the costs of facilities and their role in business operations and impacts on workforce health and safety has moved organizations to take a more strategic approach to facilities management, viewing them as assets that enable the production and delivery of goods and services. A portfolio of facilities requires continuous investment, evaluation, and strategic management if the facilities are to operate reliably, effectively, and efficiently. Because in-house facilities management divisions have outsourced many noncore facilities-related activities—including design, construction, routine operations and maintenance, and custodial services—in-house staff are now responsible for strategic planning, procurement, management, and other higher-level activities.

TABLE 1.1 Distribution of Total Assets for a Typical Private Sector Organization

Asset	Share (%)
People (salaries and benefits)	40
Facilities	23
Products/services	17
Finances	15
Miscellaneous	5

SOURCE: Brandt (1994).

Most activities contemplated in the course of an organization's strategic planning entail a facilities requirement: Space is required to house people and equipment and to ensure that operations are ongoing and efficient. The location of that space can help or hinder operations and the achievement of the organization's missions. The space may be owned or leased, depending on its function, the availability of funding, and other factors. According to the NRC (2004, p. 47),

> best practice organizations use their mission as guidance for instituting management approaches that integrate all of their resources—personnel (human capital), physical capital (facilities, inventories, vehicles, and equipment), financial capital, technologies, and information—in pursuit of a common goal.

Strategic management has been facilitated by advances in information technologies, which enable the collection and analysis of systemwide data for entire portfolios of facilities and their integration with human resources and financial data. Such integrated information gives senior executives and managers a better basis for understanding the impact of facilities-related decisions on their organization's budget, its workforce, and its missions and, in turn, allows them to make strategic decisions about when to acquire, invest in, or dispose of individual buildings.

Due to the increasingly competitive business environment, which requires optimization of facilities' functions and reliability and minimization of costs, organizations—as owners and users of facilities—have greater expectations for the performance of facilities. They are also more informed about the relationship of facilities to environmental, health, and safety issues; security threats; emergency preparedness; and continuity of operations in the post-9/11, post-Katrina era.

As a result of these changes, the role of in-house facilities divisions has evolved from tactical operations of individual buildings to facilities asset management—management of an entire portfolio of facilities.

FACILITIES ASSET MANAGEMENT

Facilities asset management has been defined as a systematic process for maintaining, upgrading, and operating physical assets cost effectively. It combines engineering principles with sound business practices and economic theory and provides tools to facilitate a more organized, logical approach to decision making (NRC, 2004). The goal of facilities asset management is to give an organization the work environments it needs to achieve its missions by optimizing available resources. This can best be done by integrating people, places, processes, and technologies to optimize the value of facilities throughout their life cycles—that is, from planning, design, and construction through operations and maintenance, renewal, and ultimately disposal (Figure 1.2).

Facilities asset management considers the entire life cycle of a facility, because 85 to 90 percent of a building's lifetime costs occur after it has been constructed (NRC, 1998). Effective life-cycle asset management takes a holistic approach that considers the interrelationships between operations and maintenance costs on the one hand and facility capital investments on the other, to help minimize total life-cycle costs and maximize asset availability and use.

A life-cycle approach allows facilities asset managers to plan strategically and to link day-to-day operations to facility investments and organizational missions. In the federal government, policies and directives have been issued for applying life-cycle costing to federal facilities investments. However, because the annual budget process considers capital and operating expenditures separately, it

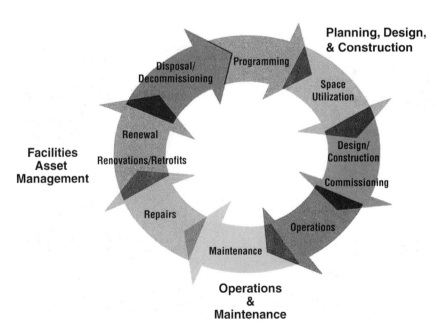

FIGURE 1.2 Facilities asset management life-cycle model. SOURCE: Adapted from APPA (2002).

discourages a total life-cycle perspective at the highest levels of decision making (NRC, 2004).

Although federal organizations are not concerned with making a profit, they also began reengineering their processes in the 1990s to improve customer service, cost effectiveness, and the outcomes of federal programs. Like private sector organizations, they began focusing on mission-essential, or core, activities and increased the outsourcing of noncore activities.

GOVERNMENT-WIDE INITIATIVES FOR MANAGEMENT REFORM

With the passage of the Government Performance and Results Act of 1993 (P.L. 103-62), Congress and two presidential administrations embarked on what has been a 15-year effort to improve the results of federal investments and programs by transforming how federal organizations (departments and agencies) produce and deliver public goods and services. Legislation and initiatives reforming policies and practices have been enacted for procurement processes,[1] accounting processes,[2] human resources,[3] and federal facilities and infrastructure.[4]

Despite these reforms, the U.S. Government Accountability Office (GAO) has designated both human capital management and federal real property management as government-wide high-risk areas. The GAO defines high-risk areas as the most important challenges facing the federal government and identifies them every 2 years at the beginning of each new Congress (GAO, 2001, 2003).

When it designated human capital "high risk," the GAO stated that "an organization's people—its human capital—are its most critical assets in managing for results" and identified four government-wide challenges:

- Strategic human capital planning and strategic alignment;
- Leadership continuity and succession planning;
- Acquiring and developing personnel whose number, skills, and deployment meet agency needs; and
- Creating results-oriented organizational cultures (GAO, 2001, p. 71).

---

[1] The Federal Acquisition and Streamlining Act of 1994 (P.L. 103-355); the Clinger-Cohen Act of 1996 (P.L. 104-208).

[2] The Government Management Reform Act of 1994 (P.L. 103-356); the Federal Financial Management Improvement Act of 1996 (P.L. 104-208).

[3] The Federal Workforce Restructuring Act of 1994 (P.L. 103-226); the Federal Activities Inventory Reform Act of 1998 (P.L. 105-270).

[4] Executive Order No. 12893, Principles for Federal Infrastructure Investments (January 26, 1994); Office of Management and Budget (OMB), Bulletin No. 94–16, Guidance on Executive Order No. 12893; OMB, Circular A–11: Part 3: "Planning, budgeting, and acquisition of capital assets"; OMB, Capital Programming Guide, a Supplement to Part 3; and Executive Order 13123, Greening the Government Through Energy Efficiency.

The President's Management Agenda (PMA) of 2001 contained a number of government-wide initiatives, one of which was "Strategic Management of Human Capital." The PMA noted that the significant downsizing of the federal workforce since 1993 had been accomplished through across-the-board staff reductions and hiring freezes rather than through targeted reductions aligned with agency missions. One result is a workforce whose skills are currently out of balance with the needs of the public it serves. The PMA further stated as follows:

> Workforce deficiencies will be exacerbated by the upcoming retirement wave of the baby boom generation, and without proper planning, the skill mix of the federal workforce will not reflect tomorrow's changing missions (PMA, 2001, p. 12).

To address this issue, the PMA called for human capital strategies to be linked to an organization's missions, vision, core values, goals, and objectives. Federal organizations must determine their core competencies and then decide whether to build internal capacity or contract for services from the private sector.

Additional legislation pertaining to human resources management in the federal government was subsequently enacted. The Chief Human Capital Officers Act of 2002 (P.L. 107-296, Title XII, Section 302) creates the position of chief human capital officer (CHCO) in the main executive departments and agencies and establishes an interagency group, the Chief Human Capital Officers Council. In Section 1305 the act also requires the Office of Personnel Management (OPM) to design a set of systems, including metrics, for assessing human capital management by agencies. The CHCO positions were created to (1) ensure that considerations of human capital and workforce management influence agency strategic planning at the highest levels; (2) create clear accountability with agencies for the responsibilities of workforce planning, leadership development, and strategic recruiting; and (3) work on metrics to gauge the progress of agencies on workforce management issues (Simpson, 2004).

Section 1412 of the Services Acquisition Reform Act of 2003 (P.L. 108-136) is intended to ensure that the federal acquisition workforce (1) adapts to fundamental changes in the nature of federal government acquisition of property and services associated with the changing roles of the federal government and (2) acquires new skills and a new perspective to enable it to contribute effectively in the changing environment of the 21st century.

In 2003, the GAO designated federal real property a high-risk area because there were significant problems with excess and underutilized property, facilities were deteriorating, data on property were unreliable, space was too expensive, and physical security was poor (GAO, 2003). In February 2004, the President issued Executive Order 13327, Federal Real Property Asset Management, which requires that each executive department and agency designate among its senior management officials a senior real property officer (SRPO) who has the education, training, and experience required to administer the necessary functions of the position for that organization. The SRPO is charged with developing and implementing an agency asset management plan that identifies and categorizes the facilities in its portfolio; sets goals and prioritizes actions for improving the operational and financial management of the portfolio; makes life-cycle cost estimations associated with those prioritized actions; and measures progress in meeting the goals. The Executive Order also establishes the Federal Real Property Council (FRPC), an interagency working group chaired by a senior executive of the OMB. The FRPC is charged with developing guiding principles for asset management, including performance measures, an inventory database, and a process for asset management planning.

In 2007, the GAO updated the status of efforts to improve both real property and human capital management. It reported that

> progress has been made [on real property], but the problems that led to the designation of federal real property as a high-risk area still exist. . . . In addition, deep-rooted obstacles, including competing stakeholder interests and legal and budgetary limitations could significantly hamper a government-wide transformation. (GAO, 2007, p. 41)

> Progress in addressing federal human capital challenges has been made since 2001, but significant opportunities remain to improve strategic human capital management to respond to current and emerging 21st century challenges. . . . The federal government now faces one of the most significant transformations of the civil service in half a century, as momentum grows toward making government-wide changes to agency pay, classification, and performance management systems. (GAO, 2007, p. 39)

DEFINING THE PROBLEM

Modern buildings and infrastructure are complex technological systems whose effective management draws on a broad range of disciplines. Facilities asset managers in both the public and private sectors are routinely called on to synthesize information from disciplines as diverse as civil and environmental engineering, materials science, government operations, economics and finance, political science, public administration, public art, design for accessibility, and conflict resolution.

At one time federal facilities management divisions primarily employed staff with technical expertise—planners, architects, engineers, project managers, real estate specialists, and other professionals—to manage their facilities portfolio and to oversee the services performed by private contractors. Now, in the face of federal workforce restructuring efforts and the ongoing retirement of experienced personnel, many facilities asset management divisions find themselves without staff having the expertise to implement a facilities asset management approach.

STATEMENT OF TASK

In response to a request from the Federal Facilities Council,[5] the National Research Council (NRC) appointed a committee of experts to undertake a study that would help to ensure effective federal facilities asset management (inclusive of property development and financial and operational functions) in the next 15 years. To this end, the committee was asked to identify and assess the following:

- Forces that will drive change in how federal buildings are planned, designed, built, operated, and managed;
- The potential impact of new and emerging technologies on processes related to facilities management;
- Organizational capabilities that federal departments and agencies will require to effectively oversee a facilities asset management program;
- Individual skills required for effective facilities asset management;
- Development strategies, processes, and training to ensure that required core competencies will be in place and sustained over time;
- Performance indicators for measuring progress in developing a workforce with the required core competencies.

THE COMMITTEE'S APPROACH

To accomplish its tasks, the committee met six times between June 2005 and July 2006. Committee members talked with representatives of federal agencies, including DoD, the Department of Energy (DOE), the General Services Administration (GSA), the U.S. Coast Guard (USCG), the National Aeronautics and Space Administration (NASA), the Smithsonian Institution, the U.S. Navy, and the U.S. Army Corps of Engineers (USACE). The committee also heard from a representative of the duPont Company.

To help in identifying organizational and individual core competencies, the committee used several previous NRC studies as a jumping-off point. It then reviewed books on facilities management, journal articles, and the programs of various professional societies and individuals. The committee's recommendations are based on a synthesis of all of these sources of information as well as on the members' own expertise and experience.

---

[5]The Federal Facilities Council (FFC) is a cooperative association of more than 20 federal agencies responsible for portfolios with many facilities. The FFC's mission is to identify and advance technologies, processes, and management practices that improve the performance of federal facilities over their entire life cycle, from planning to disposal. The FFC operates under the aegis of the NRC, the operating arm of the National Academy of Sciences, the National Academy of Engineering, and the Institute of Medicine.

PREVIOUS STUDIES OF THE NATIONAL RESEARCH COUNCIL

Three previous NRC studies examined different dimensions of the federal facilities management challenge. *Stewardship of Federal Facilities: A Proactive Strategy for Managing the Nation's Public Assets* (1998) developed a framework to facilitate strategic planning for the maintenance and repair of facilities. Its objectives were to optimize available resources to maintain the functionality and quality of federal facilities and to protect the public's investment (see Appendix C for that report's Executive Summary). *Outsourcing Management Functions for the Acquisition of Federal Facilities* (2000) provided guidance for federal agencies as they began to make decisions about outsourcing management functions such as planning, design, and construction. It also identified the core competencies federal organizations need for effective oversight of outsourced management functions while protecting the public interest (see Appendix D, which is the Executive Summary for that report). The last of the three reports, *Investments in Federal Facilities: Asset Management Strategies for the 21st Century* (2004), identified best practices of the private sector as it invests in facilities and suggested how such practices might be adapted for use in federal organizations. It proposed principles and policies for facilities investment and asset management (Appendix E contains its Executive Summary). The recommendations of these three studies were intended to result in

- Improved alignment between federal facilities portfolios and organizational missions;
- Responsible stewardship of federal facilities and federal funds;
- Savings in the life-cycle costs of facilities;
- Better use of available resources—people, facilities, and funding; and
- A collaborative environment for decision making related to federal facilities.

Several findings emerged from these studies:

- Facilities asset management in the public sector has very different objectives than such management in the private sector, even though many of the challenges are similar;
- Federal organizations should approach facilities asset management with the mindset of an owner;
- Strategic planning and decision making are needed to align the facilities portfolio with an organization's missions;
- A life-cycle approach is very important; and
- Performance measurement is a necessary ingredient of continuous improvement.

The findings of these earlier studies provided a platform for the committee as it began to address the tasks set before it and also suggested an approach for it to take.

ORGANIZATION OF THE REPORT

This report is organized into five chapters that outline the committee's approach, consensus thinking, findings, and recommendations in response to the statement of task.

Chapter 1, *Context,* provides background information about the role of facilities and infrastructure, the evolution of facilities management, and federal initiatives and issues related to the management of facilities and human resources. Chapter 1 concludes with a summary of findings from three previous NRC reports related to federal facilities management.

Chapter 2, *Forces Affecting the Federal Government: Implications for Facilities Asset Management in 2020,* categorizes and describes the diverse internal and external forces affecting the federal government and their implications for facilities and the workforce that manages them.

Chapter 3, *Core Competencies for Federal Facilities Asset Management,* defines the core competencies required for effective federal facilities asset management and describes a process that federal organizations should use to identify skills gaps.

Chapter 4, *A Comprehensive Strategy for Workforce Development,* addresses methods for developing and sustaining core competencies through a multifaceted approach to professional recruitment, education, training, and a culture of learning. Performance indicators for measuring progress in workforce development are also discussed.

Chapter 5, *Core Competencies for Federal Facilities Asset Management: Findings and Recommendations,* summarizes the committee's findings and conclusions and provides recommendations to help ensure effective federal facilities asset management through 2020 and beyond.

Appendix A contains the biographies of the committee members. Appendix B lists the briefings to the committee and interviews that took place during the study. Appendixes C, D, and E contain the executive summaries of three previous NRC reports on federal facilities management.

REFERENCES

APPA (Association of Higher Education Facilities Officers). 2002. Development of the Facility Management Profession. Alexandria, Va.: APPA.
GAO (General Accounting Office). 2001. High Risk Series: Human Capital. Washington, D.C.: GAO.
GAO (Government Accountability Office). 2003. High Risk Series: Federal Real Property. Washington, D.C.: GAO.

GAO. 2007. High Risk Series: An Update. Washington, D.C.: GAO.
NRC (National Research Council). 1998. Stewardship of Federal Facilities: A Proactive Strategy for Managing the Nation's Public Assets. Washington, D.C.: National Academy Press.
NRC. 2000. Outsourcing Management Functions for the Acquisition of Federal Facilities. Washington, D.C.: National Academy Press.
NRC. 2004. Investments in Federal Facilities: Asset Management Strategies for the 21st Century. Washington, D.C.: The National Academies Press.
PMA (President's Management Agenda). 2001. Accessed at <www.whitehouse.gov/omb/budget/fy2002/mgmt.pdf>.
Simpson, K. 2004. Testimony of Kevin Simpson, Partnership for Public Service, Before the Subcommittee on Civil Service and Agency Organization, Committee on Government Reform, House of Representatives on First Year on the Job: Chief Human Capital Officers. Available online at <http://federalworkforce.oversight.house.gov/documents>.

2

Forces Affecting the Federal Government: Implications for Facilities Asset Management in 2020

As the 21st century unfolds, geopolitical and socioeconomic trends are placing substantial pressure on the U.S. federal government. The federal presence will probably intensify both here and in some other parts of the world in response to political and economic change but diminish elsewhere. By 2020 the world's population will have increased by nearly 2 billion, and the demands on Earth's finite resources will call for heightened environmental sensitivity. Current geopolitical conditions indicate that the nation's focus on homeland security and global terrorism will continue. Rapid advances in technology, real and perceived threats to national security, changes in government paradigms, the growing national fiscal imbalance, and changes in the workforce all have tangible implications for federal facilities asset management and for the core competencies needed by the divisions set up to perform this management.

GEOPOLITICAL AND SOCIOECONOMIC TRENDS

In response to changing international political alliances and threats and domestic fiscal constraints, the portfolio of facilities that support the defense and foreign policy missions will continue to be restructured. Plans are under way to move 70,000 military troops and their dependents from Europe to the United States as part of a new defense strategy in which U.S. forces are based stateside and rapidly deployed overseas for training or military operations. This will require the closing of military bases in Europe and the upgrading of bases, including new construction, in the United States. Separately, the new round of domestic military base closures authorized in 2005 is expected to result in the closing of 22 major bases and other changes.[1] Physical security concerns are prompting the transfer of thousands of federal employees in the Washington, D.C., metropolitan area from leased facilities in commercial centers to federally owned land on which several million square feet of new space will be constructed.

The bombing of two U.S. embassies in East Africa in 1998 was the catalyst for a $16 billion, 12-year effort to provide greater security for embassy personnel. This effort will be implemented by upgrading existing buildings and constructing approximately 160 new embassies and consulates.[2]

On the domestic front, socioeconomic trends are affecting how government organizations provide services. For example, the Department of Veterans Affairs (VA) is realigning its health-care system to provide for more outpatient services and fewer long-term hospital stays. The number of hospital beds and amount of hospital space owned by the VA is being reduced, while the number of patients served through outpatient services and centers is being increased.

To respond to global and socioeconomic trends, federal facilities asset management divisions must continually evaluate whether the types, numbers, and locations of their facilities are aligned with their missions, and they must clearly have the capacity to carry out such evaluations. Strategic planning, decision making, and operations, in turn, require the capacity to identify which facilities enable or hinder the achievement of an organization's missions to formulate and evaluate alternatives for acquiring, renovating, or disposing of facilities and quantify the impacts of the various alternatives; to determine which strategies and mechanisms will be most effective in particular situations; and to effectively communicate that information to others throughout the organization, from senior executives to field office managers. Facilities asset management divisions will require staff with skills related to logistics/supply chain management, physical security, risk identification and management, and selection of the most appropriate project delivery strategies (e.g., design-build, design-bid-build).

CHANGING GOVERNMENT PARADIGMS

Contracting with private sector firms to provide goods and services that were traditionally provided by federal organizations continues to be a significant trend worldwide. A survey of a dozen national governments in the late 1990s indicated that a large majority of the respondents expected that the most successful government structure in 2010 would be one in which the national government focuses

---

[1] Four previous rounds of base closures (1988, 1991, 1993, 1995) resulted in 97 major closures, 55 major realignments, and 235 minor actions (GAO, 2005).

[2] There are more than 60,000 U.S. government employees from 35 agencies plus additional thousands of support staff at 260 diplomatic posts worldwide; most of these facilities do not meet new security standards.

on managing projects and suppliers, allowing the private sector to deliver most of the traditional public services (Economist Intelligence Unit, 1999).

Although the U.S. government has long contracted with the private sector to deliver public goods and services, the Federal Activities Inventory Reform (FAIR) Act of 1998 gave renewed emphasis to contracting out, also called outsourcing. The act broadly defines inherently governmental activities as those that are so intimately related to the public interest as to mandate performance by government employees. Inherently governmental activities include the interpretation and execution of the laws of the United States and often involve a decision pertaining to policy, prime contracts, or the commitment of government funds. The FAIR Act also states that functions that are not inherently governmental include those involving the following:

- Gathering of information for or providing advice, opinions, recommendations, or ideas to federal government officials or
- Any function that is primarily ministerial or internal in nature. Examples cited include building security, facilities operations and maintenance, and routine electrical and mechanical services.

To comply with the FAIR Act, federal organizations must annually develop lists of activities performed by government staff that in the judgment of the head of the organization are not inherently governmental functions. Such functions are classified as "commercial services." To carry out commercial services, federal organizations must use a competitive process that allows private sector firms to submit a proposal for performing the activity. Competitive sourcing, one of the five major initiatives in the President's Management Agenda (PMA), is intended to develop procedures that will expand and improve the competition between public and private organizations as envisioned in the FAIR Act.

In the foreseeable future federal agencies will continue to outsource a significant portion of services related to facilities—design, construction, operations, maintenance, and some management functions. A key question facing facilities asset management divisions is whether additional facilities-related services, including management functions, should be outsourced as well. Services that are not critical to organizational missions are likely to be considered for outsourcing whenever the risk of service failure *and* competency loss is less than the value added by an outside firm. The implication of this way of operating is significant. It demands that federal asset management divisions routinely determine which services they "own" and which they "oversee." To make this determination they must have enterprise knowledge, a profound understanding of which facilities-related services truly enable the organization's missions, and which services merely support them. To act on this knowledge, facilities asset managers must have skills in strategic planning, investment decision making, contract oversight, communication, negotiation, and risk assessment and management. They must also be able to perform detailed technical analyses of the alternatives; financial analyses of the relative costs, benefits, and cash flows of those alternatives; project management analyses; and construction management activities to implement and adjust policies, standards, and resource allocations (NRC, 2000).

As facilities asset management divisions contract out services, federal employees will need to have technical competencies related to architecture, engineering, and project management so that they can be smart buyers of services and oversee contracts. Federal facilities asset managers will still be called on to make decisions about design change orders, contract negotiations, claims and disputes, or other things that can affect safety, risk, life-cycle costs, schedule, and the community where the facilities are located. Such decisions must be based on a foundation of technical knowledge and expertise.

BUDGETARY PRESSURES

Within the federal government, decisions on investment in and management of facilities involve multiple stakeholders (Congress, the executive branch, the public), multiple decision makers (the President, Congress, the OMB, senior executives of departments and agencies), and the more than 30 facilities asset management divisions whose primary responsibility is to manage their organization's facilities portfolio. The President and Congress are responsible for providing leadership and vision, setting policies, enacting legislation, establishing regulations, and authorizing and appropriating public funds. Civil servants and political appointees in the various federal organizations are responsible for administering programs, establishing and executing processes, analyzing results, recommending initiatives, enforcing regulations, and expending public funds efficiently, effectively, and legally. In this decision-making structure, the various government entities have diverse but overlapping objectives (NRC, 2004).

In the federal budget process, funding requests typically exceed available funds. Only a relatively small proportion of the federal annual operating budget is discretionary, because the bulk of it is constrained by entitlement programs such as Social Security, Medicare, and Medicaid and by the need to maintain ongoing programs and services seen as critical or valuable. In any environment where expectations exceed resources, trade-offs must be made. Decision makers in Congress and federal departments and agencies are asked to balance the competing demands of very different programs: Funding for facilities must be weighed against funding for medical research, weapons systems, homeland security, education, and numerous other public programs.

Insufficient funding to adequately maintain the existing federal facilities portfolio is a decades-old issue. Funds appropriated by Congress to federal organizations for new construction, modernization, operations, maintenance, and management are classified as discretionary in the budget process. Projections for the federal budget through 2040 indicate that an ever-increasing proportion of federal expenditures will be needed for entitlement programs, leaving an even smaller proportion for discretionary expenditures (Figure 2.1).

This situation will place even greater demands on facilities asset managers. As federal facilities continue to age and deteriorate, the funding required for operations and maintenance will escalate exponentially. As budgetary pressures rise and as facilities portfolios become better aligned with missions, the dollars allocated to facilities will be increasingly scrutinized for their return on investment, making life-cycle planning more prevalent and visible.

As discretionary funds decrease, federal asset managers are likely to seek alternatives to budget appropriations to acquire and operate facilities. They will also be called on to dispose of excess facilities by title transfers or other means. Alternative means of acquisition such as public-private partnerships, sale-and-leaseback arrangements, real property exchanges, the sharing of facilities with other federal agencies, and title transfers will call for staff who can innovate across traditional functional lines and processes and work effectively with staff from the private sector and state and local governments. Staff will need to know about real estate finance and economics, have a profound understanding of federal regulations and procedures, and be able to recognize opportunities to accomplish organizational missions. They will need listening and negotiating skills and skills that allow them to turn an opportunity into a reality.

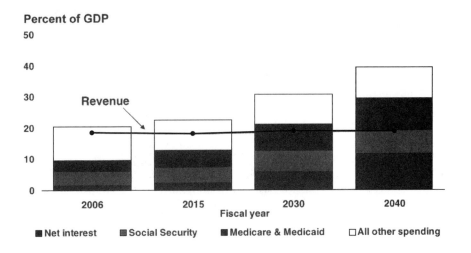

FIGURE 2.1 Spending as a share of gross domestic product. SOURCE: GAO (2006).

ADVANCES IN TECHNOLOGIES

Rapid changes in information technology (IT) have had remarkable impacts on the ways we live, work, and interact. Just as the interstate highway system provided Americans with unprecedented physical connectivity, the Internet and wireless technologies are providing virtual connectivity. Technologies already make it possible to carry out paperless transactions—for example, to obtain music without visiting a record store or to plan and execute work without the benefit of office space. The automation of routine tasks has the potential to free up workers' time to concentrate on "knowledge" work. Today, IT is an integral part of the work environment, such that an organization's information infrastructure is as important as its physical infrastructure.

Many employees of service-based private enterprises already work from home or from mobile offices, reducing the need for corporate workspace. Given current trends, it is likely that a greater portion of the federal workforce and its contractors will be working off-site, elevating the importance of digital infrastructure and knowledge management and lessening the amount of physical space required by agencies. The type of physical space required may also change, such that more meeting spaces and fewer individual workspaces will be needed.

IT is also changing the physical infrastructure and the methods for its development and management. Sensor technologies embedded in building components allow managers to gather more real-time operational data, while three- and four-dimensional models permit visualization and simulation of design and construction prior to project execution. There is every reason to believe that smart buildings will be more common by 2020, with some legacy structures having been retrofitted with infrastructures that support smart work environments. Multidimensional digital models used for building or renovating facilities will become the backbone of the facility management information system where as-built and operational data are housed. Sensors and controls will be embedded in new buildings, providing users with operational data and managerial features. Such a geospatial facility platform will allow unprecedented operational control and information access. Moreover, IT and its supporting tools will enable virtual simulations that are only being imagined today.

All of these trends are enabling a new realm of strategic and tactical decisions to be made about facilities-related impacts on the organization, the workforce, and the environment. In some cases, new technologies may be used to automate routine tasks, giving staff the time to work on more complex or strategic tasks. Facilities asset management divisions must employ staff who are knowledgeable about IT and its role in strategic decision making, design, management, and operations. They will need people who can anticipate how IT will help or hinder interaction, collaboration, and understanding among people and work units and how it will change traditional processes and the relationships among people and the organizations involved in those processes. They will need, as well, people with the leadership skills necessary to bring about changes in management, business processes, and organizational culture.

Because of downsizing and the inability to hire younger, tech-savvy workers to replace retiring federal staff, federal facilities asset management divisions generally do not have people who are able to take advantage of the opportunities presented by advances in IT or who understand their implications for workforce processes and relationships.

SUSTAINABLE DEVELOPMENT

As world population increases and the economies of China, India, and other countries continue to grow, the demand for natural resources—energy, water, building materials—may outstrip supply. Some side effects of increased development, such as air and water pollution and greenhouse gas emissions, have raised concerns about global climate change and the future quality of life on Earth. These concerns have spawned a movement toward more sustainable development.

The Bruntland Commission of the United Nations defined sustainable development as development that meets the needs of the present without compromising the ability of future generations to meet their own needs (UN, 1987). In 1996, the President's Council on Sustainable Development (PCSD) wrote that sustainable development means maintaining economic growth while producing the absolute minimum of pollution, repairing the environmental damages of the past, using far fewer nonrenewable resources, producing much less waste, and giving the whole population the opportunity to live in a pleasant and healthy environment (PCSD, 1996).

The total U.S. building stock—residential, commercial, institutional, public and private, including federal facilities—accounts for more than 40 percent of U.S. energy use as well as for significant percentages of raw materials, water, and land use. Concurrently, buildings produce 40 percent of atmospheric emissions, including greenhouse gases, and significant amounts of solid waste and wastewater (Table 2.1)

As the world's largest single owner of facilities, the federal government has a significant role to play in reducing building-related environmental impacts. Several Executive Orders[3] and the Energy Policy Act of 2005 set goals for reducing energy use and greenhouse gas emissions for federal facilities. The Energy Policy Act specifically requires a 20 percent reduction in federal building energy use by 2015 and requires the federal government as a whole to increase its use

---

[3]They include Executive Order 13123, Greening the Government Through Efficient Energy Management; Executive Order 13101, Greening the Government Through Waste Prevention, Recycling, and Federal Acquisition; Executive Order 13148, Greening the Government Through Leadership in Environmental Management; and Executive Order 14123, Strengthening Federal Environmental, Energy, and Transportation Management.

TABLE 2.1 Environmental Burdens of Total U.S. Building Stock Including Federal Facilities

Resource Use	Share of Total (%)	Pollution Emissions	Share of Total (%)
Raw materials	30	Atmospheric emissions	40
Energy use	42	Water effluents	20
Water use	25	Solid waste	25
Land (SMSAs[a])	12	Other releases	13

[a] SMSA, standard metropolitan statistical area.
SOURCE: Levin (1997).

of energy from renewable resources (e.g., hydroelectric, geothermal, wind, solar, and biomass) to 7.5 percent or more by 2013. Executive Order 14123, "Strengthening Federal Environmental, Energy, and Transportation Management," issued in January 2007, requires federal organizations to improve energy efficiency, reduce greenhouse gas emissions, and reduce water consumption intensity. It also requires organizations to ensure that new construction and major renovation of federal buildings comply with the Guiding Principles for Federal Leadership in High Performance and Sustainable Buildings and that 15 percent of the existing building inventory at the end of fiscal year 2015 incorporate the sustainable practices in the Guiding Principles.

Managing federal facilities portfolios more sustainably requires staff who understand how buildings affect and interact with the environment and how the choice of materials and design will affect both a building's life-cycle costs and the organization's long-term fiscal outlook. Sustainable management also requires individuals who can analyze data; assess the implications for energy use and water use, indoor environmental quality, and systems longevity; propose a course of action; and communicate their recommendations to decision makers and building tenants.

AGING FEDERAL WORKFORCE

As the baby boom generation approaches retirement, the federal workforce faces the possibility of workforce shortages in the near future and the loss of institutional memory and experience. Studies by the Partnership for Public Service (PPS), a nonprofit agency dedicated to making the federal government a competitive employment option, predict that retirements and resignations will result in large-scale federal turnover by 2010. Four of every ten individuals in the Senior Executive Service, which includes most career management and policy positions in the executive branch, are projected to retire by 2010 (PPS, 2006).

One potentially enduring outcome of these retirements is the loss of institutional knowledge—particularly knowledge related to the design, construction, and operation of specialized facilities such as complex military and civil works projects. As experienced personnel leave the federal government, their on-the-job knowledge of standards and criteria, unique user needs, and construction efficiencies could be lost unless efforts are made to capture it in such forms as databases, standards, decision-support tools, training manuals, or advisory councils. At best, loss of this knowledge base might result in a few project-specific inefficiencies; at worst, it could adversely affect military preparedness and national security.

ATTRACTING A NEW GENERATION OF WORKERS TO THE FEDERAL GOVERNMENT

As the bulk of today's workforce retires, the federal government will be challenged to attract a new generation of workers into its service: a tech-savvy generation born into the Information Age. This generation of workers will be accustomed to connectivity, mobility, and information availability.

The federal government has not been an employer of choice for this generation (PPS, 2006). Currently, only 3 percent of the federal workforce (military personnel excluded) is younger than 25 (PPS, 2006). A survey found that although nearly half of the college students interviewed were interested in federal service, most were unaware of federal career opportunities. Many said that "too much bureaucracy" and "low salaries" were the biggest reasons not to work for the government.

The federal hiring process, which is confusing and cumbersome, is also a barrier to recruitment. In a report to the President and the Congress, the U.S. Merit Systems Protection Board (MSPB, 2004, p. 17) said

> The Federal hiring process is often a long process. The U.S. Government Accountability Office (GAO) estimates that it takes an average of 102 days to complete all of the steps in the competitive hiring procedure (from making the request to fill the position to making the appointment). . . .The longer the process takes, the more applicant attrition is likely to increase as potential candidates accept positions with other employers. . . . The Federal hiring process is also complicated. Many applicants claim they do not understand how to apply, and this deters them from doing so.

Attracting, training, and retaining qualified staff is also an issue in the private sector because facilities asset management is evolving. A report by the Center for Construction Industry Studies (CCIS), which looked into both public and private sector firms, found as follows (CCIS, 1999, p. 30):

> It is fairly widely recognized in owner firms that the skill set required to manage and work on projects from the owner's side has changed dramatically (e.g., more "soft" skills are important; deep technical knowledge is less important).

... The issue of skill development of owner personnel is perhaps the most important difficulty facing owner firms. The vastly increased reliance on contractors has required new methods of structuring and managing projects. These new methods cannot be implemented without adequate skill and preparation of owner personnel.

An industry-wide deficit of skilled professionals means that federal organizations cannot rely on outsourcing alone as either a short- or long-term strategy for facilities asset management. To fill skill gaps and to develop and sustain core competencies and capabilities over time, federal organizations will need to implement a range of strategies, as discussed in Chapter 4.

The workplace itself is often an overlooked factor in workforce recruitment and retention, and the workforce of tomorrow will have particular expectations for the workplace environment. These expectations will likely include immediate and reliable access to information from anywhere and at any time; meeting space for communal gatherings, information exchange, and telecommuting; and tools for digital hardware, media, and communication. Thus, the configuration of the workplaces throughout the federal government will need to evolve. To support the retention and recruitment of workers, facilities managers will need to find ways to economically modify existing assets and design new space. An "information and control infrastructure" will become one of the central components of the facilities portfolio, and its reliability will be essential. Facilities asset managers will need to coordinate across traditional functional lines to bring together information technology groups and other tenants as facilities are modernized, expanded, and operated.

A NEW PARADIGM IS ESSENTIAL

By 2020, the divisions that manage federal facilities will require professionals with a multifaceted, wide-ranging set of knowledge, abilities, and skills if they are to effectively support the missions of their organizations. Their staff must collectively be able to

- Identify which facilities enable or hinder the achievement of the organization's missions.
- Formulate and evaluate alternatives for acquiring, renovating, or disposing of facilities.
- Use information technology effectively for decision support.
- Identify and quantify the potential consequences of decisions.
- Develop and execute strategic plans to align current facilities portfolios and available resources to missions.
- Act as smart buyers of services and oversee contracts.

- Work with staff from other agencies, state and local governments, and the private sector.
- Assess and manage risk.
- Determine which strategies and mechanisms will be most effective in particular situations.
- Communicate information effectively to others inside and outside their organization.
- Innovate across traditional functional lines and processes.

To fulfill these responsibilities, the staff of facilities management divisions will need a range of technical, behavioral, and business skills and enterprise knowledge to effect changes in management, business processes, and the organizational culture.

Although federal organizations face many challenges in developing such a workforce, they also have a significant opportunity: As the current workforce retires and as technologies become increasingly important in decision support and daily operations, their asset management divisions can redefine their core competencies and then hire, train, and equip a workforce that has the knowledge, skills, and abilities required to move forward.

To meet current and future challenges, federal facilities asset management divisions must continue to evolve and do so quickly. The director of such a division and the senior real property officer should be involved in the organization's strategic planning and executive-level decision making, which requires a macroscopic, strategic view of the assets being managed. Moreover, the staff of these divisions must have the appropriate education and skills and be organized into a structure that enables achievement of organizational missions by linking day-to-day operations and facility investments.

Thus a new paradigm is essential. Indeed, one of the more complex and exciting challenges confronting federal asset management divisions is the formulation of strategies to support the migration of their organizations from the workplace of today to the workplace of tomorrow. The pending retirements of significant numbers of federal employees, coupled with new technologies and an increased appreciation for the role of facilities in achieving missions, present an opportunity to transform facilities asset management divisions and grow a new generation of facilities managers. The core competencies—essential areas of expertise and the skills base—required by facilities asset management divisions are the subject of Chapter 3.

REFERENCES

CCIS (Center for Construction Industry Studies). 1999. Owner/Contractor Organizational Changes, Phase II Report. Austin: University of Texas Press.

Economist Intelligence Unit and Andersen Consulting. 1999. Vision 2010: Forging Tomorrow's Public-Private Partnerships. New York.

GAO (Government Accountability Office). 2005. Military Base Closures: Observations on Prior and Current BRAC Rounds. Washington, D.C.: GAO.

GAO. 2006. The Nation's Long-Term Fiscal Outlook. Washington, D.C.: GAO.

Levin, H. 1997. Systematic Evaluation and Assessment of Building Environmental Performance. Second International Conference on Buildings and the Environment, CIB TG8, Environmental Assessment of Buildings, June 9-12, Paris. Rotterdam, The Netherlands: International Council for Research and Innovation in Building and Construction.

MSPB (U.S. Merit Systems Protection Board). 2004. Managing Federal Recruitment: Issues, Insights, and Illustrations. Washington, D.C.: MSPB.

PCSD (President's Council on Sustainable Development). 1996. Sustainable America: A New Consensus for Prosperity, Opportunity and a Healthy Environment for the Future. Available at <http://clinton2.nara.gov/PCSD/Publications/>. Accessed November 21, 2006.

PPS (Partnership for Public Service). 2006. Back to School: Rethinking Federal Recruiting on College Campuses. Washington, D.C.: Partnership for Public Service.

UN (United Nations), World Commission on Environment and Development. 1987. Our Common Future. New York: Oxford University Press.

3

Core Competencies for Federal Facilities Asset Management

In 1990, C.K. Prahalad and Gary Hamel introduced the concept of core competencies as a significant component of effective management. They defined "core competency" as an area of specialized expertise that is the result of harmonizing complex streams of technology and work activity (Prahalad and Hamel, 1990). As the concept has evolved, so has its definition, and it is now variously thought of as "the sum of learning across individual skill sets and individual organizational units" (Hamel and Prahalad, 1994); "aggregates of capabilities, where synergy is created that has sustainable value and broad applicability" (Gallon et al., 1995); and "a combination of complementary skills and knowledge bases embedded in a group or team that results in the ability to execute one or more critical processes to a world class standard" (Coyne et al., 1997).

In this study, the committee has defined core competencies as an organization's essential areas of expertise and the skills base required for achieving its missions. The operational unit for which core competencies are being defined here is the facilities asset management division. The mission is specific to the organization (department or agency) for which the facilities asset management division provides support. Within the division, a facilities asset manager's competencies include any skill, knowledge, behavior, or other personal characteristic that is essential for performing the required functions.

To help identify the core competencies required by such divisions through 2020 and beyond, the committee (1) reviewed relevant articles, publications, and reports; (2) conducted interviews and discussions with representatives of facility management professional organizations and reviewed organizational information; (3) reviewed higher education and professional development programs; and (4) interviewed federal staff. The information it gathered is summarized below.

FACILITIES MANAGEMENT COMPETENCIES LITERATURE

The committee's literature search centered on facilities management at a strategic level, where the practitioners would be considered as having professional standing. A common theme is that facilities asset management is evolving as a business management discipline and will not remain rooted in operational and cost-centric issues (Price, 2004; Rodgers, 2004).

Markus and Cameron (2002) characterized the competencies of facilities management as follows:

- *Operational maintenance.* A technical function concerned with maintaining the practical utility of the physical infrastructure to ensure that it supports the core activity of an organization.
- *Financial control.* An economic function concerned with ensuring the efficient use of physical resources by controlling costs.
- *Change management.* A strategic function concerned with the forward planning of physical infrastructure resources to support organizational development and reduce risk.
- *User interfacing.* A social function concerned with ensuring that the physical infrastructure of work meets the legitimate needs of users within their organizational role.
- *Advocacy.* A professional function that includes social responsibility for people in the workplace.

Others predict that facilities asset management divisions will be staffed by knowledge workers, who are able to assimilate business, people, processes, and property knowledge in order to develop innovative facilities solutions (Nutt and McLennan, 2000). In this model, a resource-based approach described as "four trails to the future" provides a framework for evaluating the tensions that occur at the interfaces between the trails (Figure 3.1).

A report by the Center for Construction Industry Studies (CCIS) found that the categories of knowledge and skills needed to manage work on projects from the owner's side have changed dramatically (CCIS, 1999) (Table 3.1).

Additional competencies identified in the literature can be categorized as follows:

- Strategic management and business knowledge (Lopes, 1996; Hinks, 2001; Cotts and Rondeau, 2004).
- Service management (Rodgers, 2004).
- Human resources and people management (McGregor, 2000).
- Performance measurement (PCA, 2000).
- Procurement strategies (Price and Akhlaghi, 1999).

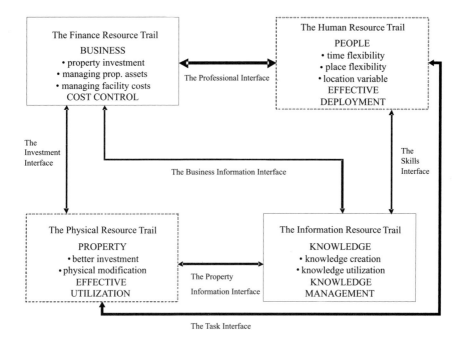

FIGURE 3.1 The resource model of strategic facilities management. SOURCE: Nutt (2000).

Another paradigm for facilities asset management divisions focuses on the need to be informed by clients to be able to deliver customer satisfaction and achieve best value for the resources expended. In *Total Facilities Management* (Atkins and Brooks, 2005), the authors list the key attributes for facilities managers as follows:

- Understanding the organization, its culture, and its customers and their needs.
- Understanding and specifying service requirements and targets.
- Brokering services among stakeholders.
- Managing risk.
- Managing contractors and monitoring their performance.
- Benchmarking the performance of outsourced service(s).
- Developing, with service providers, strategies for delivering service(s).
- Understanding strategic planning.
- Safeguarding public funds, where relevant.

TABLE 3.1 Knowledge and Skills Required for Facilities Asset Management Divisions

Category of Knowledge	Skill Set
Business	Writing and managing contracts
	Negotiating
	Managing budgets and schedules
Communication	Coordinating/liaising
	Managing conflict
	Cultivating broad network of relationships
Influence	Mentoring
	Motivating
	Managing change
Managerial	Building teams
	Delegating
	Gaining political awareness/seeing the big picture
Problem solving	Continuously analyzing options/innovations
	Planning
	Considering all sides of issues, risk management
Technical	Understanding entire construction process
	Having multidisciplined knowledge of several areas of engineering
	Having information technology skills

SOURCE: Adapted from CCIS (1999).

- Developing in-house skills through education, training, and continuing professional development.

FACILITIES MANAGEMENT COMPETENCIES IDENTIFIED BY PROFESSIONAL ASSOCIATIONS

The committee also reviewed materials related to core competencies for facilities asset management published by professional associations, including the International Facility Management Association (IFMA), the Association of Higher Education Facilities Officers (APPA),[1] the Building Owners and Managers Association (BOMA), the Institute of Real Estate Management (IREM),

---

[1] This organization was formerly called the Association of Physical Plant Administrators (APPA). When the name was changed to Association of Higher Education Facilities Officers, the acronym APPA was kept. See Web site at <www.appa.org>.

the Association of Facility Engineers (AFE), the British Institute of Facilities Management (BIFM), the Royal Institute of Chartered Surveyors (RICS), and the Facility Management Association of Australia (FMAA). Table 3.2 summarizes the definitions of facilities management from three international organizations and their related core competencies.

TABLE 3.2 Facilities Asset Management Competencies

IFMA	BIFM	FMAA
Facility management is "a profession that encompasses multiple disciplines to ensure functionality of the work environment by integrating people, place, processes and technology."	Facility management is "the integration of multidisciplinary activities within the built environment and the management of their impact upon people and the workplace."	Facility management is "a business practice that optimizes people, process, assets, and the work environment to support the delivery of an organization's business objectives."

Core competencies
  Leadership and management
  Human and environmental factors
  Planning and project management
  Operations and maintenance
  Finance
  Real estate
  Communication
  Quality assessment and innovation
  Technology
	Core competencies
	  Understanding business organizations
	  Managing people
	  Managing premises
	  Managing services
	  Managing the work environment
	  Managing resources
		Competency standards
		  Use organizational understanding to manage facilities
		  Develop strategic facility response
		  Manage risk
		  Manage facility portfolio
		  Improve facility performance
		  Manage the delivery of services
		  Manage projects
		  Manage financial performance
		  Arrange and implement procurement /sourcing
		  Facilitate communication
		  Manage workplace relationships
		  Manage change

Focus
  Multidisciplinary profession
  Functionality of the work environment
  Integration of resources (people, place, and technology) and processes
	Focus
	  Multidisciplinary activities
	  Relationship between built environment and workplace
	  Relationship between people and the workplace
		Focus
		  Driven by business objectives
		  Coordinating management function
		  Supporting business function

SOURCE: Then (2004).

Interviews with representatives of other professional facilities management organizations yielded similar insights into competencies. Looking at similar data, Warren and Heng (2005, p. 12) concluded that such similarities

> [bode] well for the continuing growth of the [facilities management] profession globally as a single discipline and for its recognition as a valuable contributor to organization's profitability [sic]. The skills also reveal the very broad basis of facilities management practice, emphasizing the need for practitioners to obtain specific qualifications to enhance their knowledge and ability to perform in this competitive business discipline.

REVIEW OF EDUCATIONAL AND PROFESSIONAL DEVELOPMENT PROGRAMS

The committee also identified colleges and universities offering degrees and professional certifications in facilities management. The research included U.S. institutions and some international universities recognized by IFMA. Not all universities identified as providing facilities management courses were evaluated and included in the results.

The committee's research into educational programs in this country was conducted by reviewing information made available on the Web sites of facilities management programs and in academic syllabi. Where Web sites did not provide adequate course details, follow-up interviews were conducted. From this research, the committee developed a matrix summarizing competencies by institution and program.

Table 3.3 lists core competency courses currently available in facilities management educational programs. The programs include those that award B.S. and M.S. degrees and certificate programs. Check marks indicate that the educational program offers a course whose title generally reflects the listed competency or that the primary content of the course is similar to the corresponding competency. An argument can be made that the educational programs cover more competencies than are identified in the table. Some programs may in fact cover most or all of the listed competencies. For simplicity, however, the table is limited to primary course titles and objectives.

IDEAS OF FEDERAL AGENCIESS ON COMPETENCIES REQUIRED FOR FACILITIES MANAGEMENT

The committee also focused on the culture and needs of existing federal facilities asset management divisions. Briefings and interviews were held with representatives from the Department of Energy (DOE) and the Department of Defense (DoD), the General Services Administration (GSA), the U.S. Coast Guard (USCG), the Naval Facilities Engineering Command (NAVFAC), and the

CORE COMPETENCIES FOR FEDERAL FACILITIES ASSET MANAGEMENT

TABLE 3.3 Core Competencies Based on Current Facility Management Educational Programs

	FM Intro/Fundamentals/Theories	General Administration	Operations & Maintenance	Planning & Design	Project Management/Const. Mgmt.	Construction Documents	Cost Estimating/Quantity Takeoffs	Contracting and Procurement	Technology - IT	Real Estate	Strategic Planning/Organization	Risk/Business Continuity[a]	Communication	Leadership/Ind. Effectiveness	Building Systems/Const. Materials[b]	Energy Mgmt. and Utilities	Finance and Accounting	Human & Environmental Factors	Environmental Issues/IAQ	Quality, Innovation, Performance Measurement	Asset Management	Ethics/Law	Leasing	Marketing Management	Future Issues/Trends in FM	Code Compliance/Safety	Office and Furniture Design	Space Planning/Interior Design	Human Resource Management
Associations																													
IFMA	•		•	•	•						•	•	•			•	•	•		•									
APPA		•	•	•							•			•		•	•							•	•	•	•		
BOMI	•		•	•					•	•		•		•		•	•	•	•			•	•	•	•				
RICS	•		•	•	•	•	•	•	•	•	•					•	•	•	•	•	•	•	•	•	•	•		•	•
IREM			•							•		•					•		•				•	•					•
B.S. degree programs																													
Brigham Young	•		•		•	•	•			•				•			•	•			•			•			•	•	•
Ferris State	•			•	•					•							•				•							•	
Cornell (also M.S. deg.)		•								•				•			•		•			•	•	•					•
Wentworth Institute of Tech.	•	•	•	•							•			•	•	•		•											•
North Dakota State University	•											•				•										•		•	
M.S. degree programs																													
Georgia Tech	•		•	•	•								•					•	•	•					•	•			
Texas A&M	•	•	•	•		•									•														
Arizona State University - FMRI	•									•					•	•	•			•	•								
Certificate programs																													
George Mason	•	•	•	•	•				•	•			•			•			•	•		•			•	•			
UC Berkeley	•		•	•	•					•	•	•	•	•	•	•													
UC Irvine	•		•		•		•			•	•				•				•			•						•	•
University of Washington	•		•	•	•		•	•			•				•			•			•	•						•	•
International programs																													
Hong Kong Polytechnic	•		•	•	•					•	•					•	•	•		•									
FHS Kufstein Bild. (Austria)	•	•	•											•			•			•				•					
Hanze University (Netherlands)	•	•									•						•				•				•			•	
United Kingdom (Various)	Several universities offer FM specialties within the Building Surveying curriculum (Leeds, UWE, Reading, Manchester, University College, etc.)																												
Textbooks																													
APPA FM Manual[c]	•	•	•	•			•			•	•	•	•	•	•	•	•									•	•		
The FM Handbook[d]	•	•	•	•	•			•	•	•	•				•			•		•		•		•	•		-		

NOTE: Bullets indicate only core competency course title and primary content. Additional competencies may be covered in each course but are not designated in the matrix. Other topics include engineering systems processes, sustainability, CADD, building safety, EH&S, benchmarking, cost estimating, property development, public FM.

[a]Includes disaster recovery and emergency preparedness planning.

[b]May include design, evaluation, operation and maintenance of HVAC, MEP, FL&S, security, conveying, structural, and waterproofing systems. IFMA-recognized programs.

[c]APPA, Facilities Management: A Manual for Plant Administration, 2nd Ed., 1989.

[d]D. Cotts, The Facility Management Handbook, New York: AMACOM, 1999.

U.S. Army Corps of Engineers (USACE). Emerging from these discussions were three issues of critical importance:

- Defining the skill sets needed to build leaders.
- Developing a flexible model for facilities asset management divisions to meet differing organizational missions (one size does not fit all).
- Identifying requisite business skills and general competencies.

Further discussions revolved around the future of facilities management as a profession and the skills that would be possessed by an ideal facilities asset manager. When comments amassed in the interviews were carefully considered, they seemed to suggest that a good federal facilities asset manager would have the following attributes:

- A financial perspective and understanding the time value of money, facilities, and other resources.
- Leadership abilities and initiative and ability to manage risk and measure performance.
- Understanding of customer requirements.
- Good decision-making capability and ability to plan and to focus on outcomes.
- Understanding of budgeting and allocation of resources.
- People and communication skills.
- A technical background in, or understanding of, building systems.
- Ability to translate technical issues into business requirements.
- Ability to assess and value real estate.
- Technological savvy.
- Program and project management skills.

One of the more established and comprehensive federal government educational programs evaluated was prepared for the facilities engineering (FE) career field by the Defense Acquisition University (DAU). The DAU provides assignment-specific continuing education courses for military and civilian acquisition personnel within the DoD. Its mission is to provide federal workers with the tools and knowledge to make smart business decisions.

The DAU's development of the FE courses involved the identification of competencies, performance outcomes, learning objectives, and assessments. The learning objectives of the intermediate level, FE 201, include these:

- Program management,
- Contracting,
- Design and construction,
- Financial laws, regulations, and procedures,

- Real estate acquisition, management, and disposal,
- Environmental requirements,
- Comprehensive and project planning,
- Managing sustainment,
- Restoration and modernization, and
- Contingency engineering processes.

The advanced course, FE 301 (not yet implemented) (Baecher et al., 2005), identifies competencies required of federal facilities asset managers, who must be able to

- Contrast approaches to FE and the facilities life cycle across the military services, both in concept and in practice.
- Assess strategies for real estate acquisition and disposal.
- Defend a funding and resource strategy to support FE-related mission requirements.
- Judge the adequacy of specific program documentation, including documentation for acquisition baseline, risk management plan, budget estimates, acquisition plan, total ownership cost for the FE life cycle, antiterrorism force protection, including commercial best practices to support the facilities life cycle.
- Distinguish among common cost, schedule, and quality risks; select appropriate risk handling options and metrics.
- Defend an acquisition plan that supports an FE mission requirement.
- Adapt non-DoD government and industry best practices for inclusion in DoD FE.
- Evaluate total ownership cost (TOC) for the FE life cycle in an acquisition strategy.
- Develop effective controls to manage variations from baseline plans for cost, schedule, and quality in facilities life-cycle activities.
- Evaluate responses to external interventions in facilities life-cycle activities, including those by local stakeholders, congressional delegations, the public, and others.

LEADERSHIP SKILLS

One of the critical issues identified by federal staff was defining the skill sets needed to build leaders. Building leadership skills has also been identified as a key ingredient for a successful transformation of all federal programs and practices (Figure 3.2).

Testifying before Congress in 2005, David Walker, comptroller of the United States, identified committed, persistent, consistent leadership as a transformational factor in meeting 21st century challenges. He also said that leadership

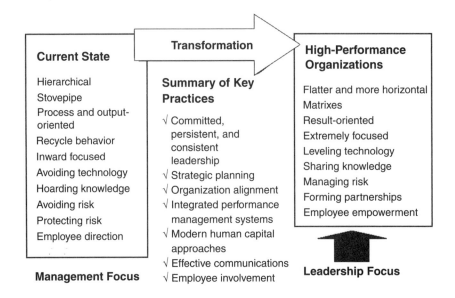

FIGURE 3.2 Cultural change and key practices for meeting 21st century challenges. SOURCE: Adapted from GAO (2005).

must set the direction, pace, and tone for the transformation and should provide sustained and focused attention over the long term (GAO, 2005).

The Office of Personnel Management (OPM) has developed a set of executive core qualifications (ECQs) that are required for entry to the Senior Executive Service. The ECQs are used by many departments and agencies in selecting personnel for management and executive positions, then managing their performance and developing their leadership. The ECQs are intended to define the competencies needed to build a federal corporate culture that motivates for results, serves customers, and builds successful teams and coalitions within and outside the organization (OPM, 2007).

The ECQs defined by OPM include these:

- *Leading change.* The ability to bring about strategic change, both within and outside the organization, to meet organizational goals, and to establish an organizational vision and implement it in a continuously changing environment.
- *Leading people.* The ability to guide people to meet the organization's

vision, mission, and goals and to provide an inclusive workplace that fosters professional development, facilitates cooperation and teamwork, and supports constructive resolution of conflicts.
- *Results driven.* The ability to meet organizational goals and customer expectations and to make decisions that produce high-quality results by applying technical knowledge, analyzing problems, and calculating risks.
- *Business acumen.* The ability to manage human, financial, and information resources strategically.
- *Building coalitions.* The ability to build coalitions internally and with other federal agencies, state and local governments, nonprofit and private sector entities, foreign governments, or international organizations to achieve common goals.

To fulfill the core qualifications, executives must have a set of fundamental competencies as well as competencies in particular areas (Table 3.4)

The committee also reviewed leadership competency models developed for the U.S. Army, a military organization, and the National Aeronautics and Space Administration (NASA), a civilian agency. Technical Report 1148, *Competency*

TABLE 3.4 OPM's Fundamental Competencies and Executive Core Qualifications

Fundamental Competency	Leading Change	Leading People	Results Driven	Business Acumen	Building Coalitions
Interpersonal skills	Creativity and innovation	Conflict management	Accountability	Financial management	Partnering
Oral communication	External awareness	Leveraging diversity	Customer service	Human capital management	Political savvy
Integrity/ honesty	Flexibility	Developing others	Decisiveness	Technology management	Influencing, negotiating
Written communication	Resilience	Team building	Entrepreneurship		
Continual learning	Strategic thinking		Problem solving		
Public service motivation	Vision		Technical credibility		

Based Future Leadership Requirements, describes a competency framework for leadership requirements for the future Army (Horey, 2004). The authors defined a competency as "a cluster of knowledges, skills, abilities, and other characteristics (KSAOs) that underlies effective individual behavior leading to organizational success." On page 2, Horey and his colleagues noted that "competencies, if well defined and comprehensive, should provide individuals and organizational processes with the roadmap that identifies successful performance of leadership duties and responsibilities." They compared leadership models from the USCG, the Army (USA), the Marine Corps (USMC), the Air Force (USAF), the Navy (USN), and the OPM (EQC) and synthesized their findings as shown in Table 3.5.

The constructs that appear in three or more frameworks of the various organizations are these:

- Performing/executing/accomplishing the mission,
- Vision/planning/preparing,
- Problem solving/decision making,
- Human resources management,
- Process/continuous improvement,
- Motivating/leading people,
- Influencing/negotiating,
- Communicating,
- Teamwork/team building,
- Building/developing partnerships,
- Interpersonal skills,
- Accountability/service motivation,
- Values,
- Learning (including components of adaptability, flexibility, awareness),
- Technical proficiency,
- Driving transformation/leading change,
- Strategic thinking,
- Diversity management,
- Mentoring/developing people, and
- Physical/health/endurance.

Based on this research, Horey et al. developed a core leadership competency framework for the Army with eight core competencies, as follows (p. viii):

- Leading others to success.
- Exemplifying sound values and behaviors.
- Vitalizing a positive climate.
- Ensuring a shared understanding.
- Reinforcing growth in others.

TABLE 3.5 Comparison of Service Framework Constructs for Leadership Competencies

USCG	USA	USMC	USAF	USN	ECQ
Performance	Executing: operations	Accomplish tasks	Driving execution	Accomplishing mission	Results driven
Influencing others	Influencing		Influencing and negotiating	Influencing and negotiating	Influencing and negotiating
Working with others	Motivating		Leading people and teams	Leading people; working with people	Leading people
Management and process improvement	Improving	Initiative	Driving continuous improvement	Continuous improvement	
Effective communication	Communicating	Keep Marines informed	Fostering effective communication	Oral/written communication	Oral/written communication
Develop vision and implement	Planning; preparing		Creating/demonstrating vision	Vision	Vision
Decision making; problem solving	Conceptual; DM	Make sound decisions	Exercising sound judgment	Problem solving; decisive	Problem solving; decisive
Group dynamics	Building; developing	Train team	Fostering teamwork	Team building	Team building
Self-awareness; learning	Learning	Know self and improve	Assessing self		Continual learning
Technical proficiency	Technical	Technical proficiency		Technical credibility	Technical credibility
Aligning values	7 values LDRSHIP	7 values	Leading by example	Integrity	Integrity, honesty

NOTE: DM, diversity management; LDRSHIP, loyalty, duty, respect, selfless service, honor, integrity, and personal courage.
SOURCE: Horey et al. (2004).

- Arming self to lead.
- Guiding successful outcomes.
- Extending influence.

NASA has also developed a leadership model that defines dimensions—highest level elements—for senior leaders:

- Personal effectiveness,
- Competency in particular disciplines,
- Managing information and knowledge,
- Business acumen,
- Leading and managing others, and
- Working internationally.

Similar to OPM's EQC model, the NASA model identifies a series of com-

TABLE 3.6 NASA's Competencies for Senior Leaders

Personal Effectiveness	Discipline Competency	Managing Information and Knowledge	Business Acumen	Leading and Managing Others	Working Internationally
Cognitive skills	Understanding of discipline	Awareness and use of information technology	Organizational culture	Leading and managing change	Cross-cultural relationships
Relating to others	Safety	Knowledge management	Organizational strategy	Leading and managing organizations	International
Personal capabilities and characteristics	Maintain credibility		Internal and external awareness	Leading and managing work	
	Communication and advocacy		Business development		
			Business management		
			Customer, stakeholder, and partner relationships		

SOURCE: NASA (2007).

petencies required to fulfill each trait. Competencies are defined as measurable skills, knowledge, or personal characteristics that have been demonstrated to be essential to effective leadership in NASA (Table 3.6).

SUMMARY

The committee identified several consistent themes from its research:

- Facilities asset management encompasses people, places, technologies, processes, and knowledge.
- Facilities asset management divisions must integrate functions and resources to align them with the mission of the organization.
- The operating environment for facilities asset management is dynamic and requires the capacity to innovate.
- The skills base for facilities asset management divisions should include an appropriate balance of technical, business, and behavioral capabilities and enterprise knowledge.
- Leadership capabilities require wide-ranging management skills, behaviors, and values that allow an individual to make effective decisions, influence others, and lead them to success.

Of all the stakeholders involved in funding, programming, designing, constructing, operating, and maintaining facilities, the facilities asset management division is the only one involved throughout all phases, serving as the dominant player throughout the operations and maintenance phase, which accounts for 90 percent of the life cycle and facility investments (Figure 3.3).

Facilities asset management divisions are responsible for a wide range of functions, all using different technologies and involving different operational units (construction, design, energy, engineering, electrical, environmental, fire, forestry, historic preservation, IT, janitorial, landscaping, maintenance, mechanical, parking, power, safety, sanitary, security, space, traffic, and others). This diversity means that such divisions must serve as "connected integrators" of diverse stakeholders, functions, and services across the life cycle of facilities.

Facilities asset management divisions must also have people with capacity for strategic thinking and planning that leads to alignment of the facilities portfolio with the organization's missions. In a dynamic operating environment, these divisions must be able to innovate across traditional functional lines and processes to address changing requirements and to take advantage of opportunities as they arise.

REQUIRED CORE COMPETENCIES

Based on its literature review, briefings, interviews, current geopolitical and socioeconomic trends, and the experience and knowledge of its members, the

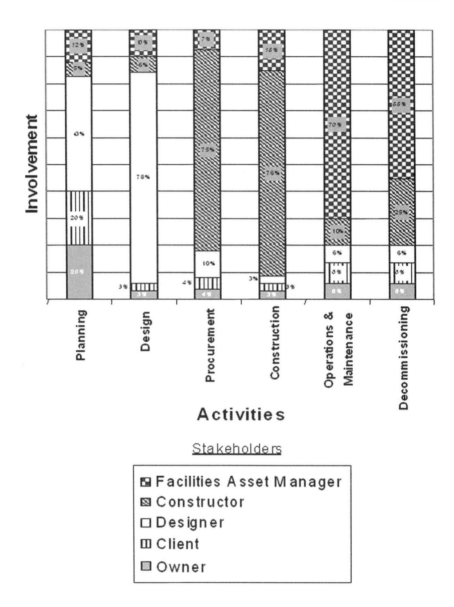

FIGURE 3.3 Stakeholders' involvement in various phases of a facility's life cycle, as estimated by the committee.

committee concluded that three areas of expertise are essential for federal facilities asset management divisions through 2020 and beyond, as follows:

- Integrating people, processes, places, and technologies by using a life-cycle approach to facilities asset management;
- Aligning the facilities portfolio with organizational missions and available resources; and
- Innovating across traditional functional lines and processes to address changing requirements and opportunities.

The required skills base includes a balance of technical, business, and behavioral capabilities and enterprise knowledge. Technical capabilities such as engineering and architecture provide the foundation and expertise for life-cycle asset management and include knowledge of facilities-related systems, their operation and maintenance; acquisition and project management processes; regulations and procedures; and technologies and analytical capabilities. Business capabilities focus on strategic planning and resource management to advance an organization's missions. Behavioral capabilities involve the leadership, communication, negotiation, and change management skills required to integrate functions, people, and processes across traditional lines and the capacity to innovate within a dynamic operating environment. Enterprise knowledge includes an understanding of the facilities portfolio and how to align it with the organization's missions; an understanding of the organization's culture, policy framework, and financial constraints; agency inter- and intradependencies; and the workforce's capabilities and skills. Facilities asset managers will require technical skills, enterprise knowledge, behaviors, and other personal characteristics that allow them to work in a team-oriented environment and to support achievement of a life-cycle approach for facilities asset management. Together, the three essential areas of expertise and the skills base make up the core competencies for federal facilities asset management divisions through 2020 and beyond.

IDENTIFYING CORE COMPETENCIES FOR FACILITIES ASSET MANAGEMENT DIVISIONS

To develop its essential areas of expertise and skills base, each facilities asset management division must first identify the functions it needs to perform in support of its organization's current and future missions. For example, the facilities asset management division within DOE may focus on operations, maintenance, cleanup, and disposal of facilities; that same division within the Office of the Architect of the Capitol may focus on space management and historic preservation issues; and the Public Buildings Service of the GSA may prioritize real estate, leasing, and contracts/acquisition/disposal. Some similarities across facilities asset management divisions will surface, but the unique aspects of each organiza-

TABLE 3.7 Examples of Functions and Skills That Might Be Required to Support an Organization's Missions

Technical	Business	Behavioral	Enterprise Knowledge
Operations and maintenance	Strategic planning	Leadership	Mission
Planning and design	Asset management	Teamwork/team building	Vision
Building systems	Finance and accounting	Interpersonal relationships	Strategic direction
Project management	Contract monitoring	Mentoring/coaching	Values
Construction	Procurement	Negotiating	Culture/trust
Code compliance	Real estate	Critical thinking	Systems
Cost estimating	Acquisition and leasing	Communication	Processes
Space planning	Business lexicon	Change management	Resource allocation
Environmental health and safety	Risk management	Quality and innovation	
Energy management	Contingency planning	Future issues/trending	
FM technology	Ethics/law	Performance measurement	
Sustainability	Marketing	Benchmarking	
Commissioning	Human resources		
Security	Professional development		
Life-cycle analyses	Organizational planning		

NOTE: FM, facilities management.

tion will become apparent only after a careful analysis of its current and future functional requirements. Examples of the types of functions and skills that may be required are shown in Table 3.7.

Once the required functions are identified, it should be determined if the division is structured to enable effective life-cycle facilities asset management or if some reorganization is required to better support a life-cycle approach.

The next step is to look at the individuals who will collectively be responsible for facilities asset management and as a group will need the knowledge, skills, abilities, and behaviors to fulfill the required core competencies. Facilities asset management divisions will need to ask and answer the following questions:

- What skills are possessed by the facilities asset managers who currently work in the facilities asset management division? Are these skills accessible/available in sufficient quantity for effective facilities asset management? What skills are found in people who are eligible to retire within 5 years?
- Is there a gap between the current skills base and the skills required for effective facilities asset management in 2020?
- What steps should be taken to close the skill gaps? What skills and capabilities are best acquired through new hires? Which by contracting out?

Which by training of current staff? Which through a recognized career path?

The steps to be taken to close skills gaps will depend on the skills at issue and will require a comprehensive workforce development strategy. For example, technical skills might best be acquired by hiring recent college graduates with degrees in facilities management, architecture, engineering, business administration, public administration, or related fields. In contrast, business skills might be strengthened by hiring experienced professionals from private companies, which value individuals with financial and risk management skills. Or, to acquire leadership skills, experienced military engineering officers who have undergone leadership training might be hired. Enterprise knowledge might be best developed by providing an in-house career path that fosters the professional development of current and new employees through a variety of strategies over time. A comprehensive strategy for workforce development is the subject of Chapter 4.

REFERENCES

Atkins, B., and A. Brooks. 2005. Total Facilities Management. Oxford, England: Blackwell Publishing.
Baecher, G., J. Dean, et al. 2005. Competencies—FE 301 Facilities Engineering Level III. Draft. Fort Belvoir, Va.: Defense Acquisition University.
CCIS (Center for Construction Industry Studies). 1999. Owner/Contractor Organizational Changes, Phase II Report. Austin: University of Texas Press.
Cotts, D., and E.P. Rondeau. 2004. The Facility Manager's Guide to Finance and Budgeting. New York: AMACOM.
Coyne, K.P., S.J.D, Hall, and P.G. Gorman. 1997. "Is your core competence a mirage?" McKinsey Quarterly 1: 40-54.
Gallon, M., H.M. Stillman, and D. Coates. 1995. "Putting core competency thinking into practice." Research Technology Management 38 (3): 20-29.
GAO (Government Accountability Office). 2005. 21st Century Challenges: Transforming Government to Meet Current and Emerging Challenges. Testimony of David Walker. Available online at <www.gao.gov/newitems/d05830T.pdf>.
Hamel, G., and C.K. Prahalad. 1994. Competing for the Future. Cambridge, Mass.: Harvard Business School Press.
Hinks, J. 2001. "All FM is non-core, but is some FM less core than others? A discussion on realizing the strategic potential of FM." FMLink white paper.
Horey, J. 2004. Technical Report 1148, Competency Based Future Leadership Requirements. Arlington, Va.: U.S. Army Research Institute for the Behavioral and Social Sciences.
Lopes, J.L.R. 1996. "Corporate real estate management features." Facilities 14 (7/8).
Markus, T.A., and D. Cameron. 2002. The Words Between the Spaces: Buildings and Language. London, England: Routledge.
McGregor, W. 2000. "Preparing for an uncertain future," Facilities 18 (10/11/12).
Nutt, B., and P. McLennan. 2000. Facilities Management: A Strategy for Success. Oxford, England: Chandos Publishing.
Nutt, B. 2000. "Four competing futures for facility management." Facilities 18 (3/4).

OPM (Office of Personnel Management). 2007. Ensuring the federal government has an effective civilian workforce. Available online at <http://www.opm.gov/ses/qualify.asp>.

Prahalad, C.K., and Gary Hamel. 1990. "The core competence of the corporation." Harvard Business Review 68 (3): 79-91.

Price, I. 2004. "Business critical FM." Facilities 22 (13/14).

Price, I., and Akhlaghi, F. 1999. "New patterns in facilities management: Industry best practice and new organisational theory." Facilities 22 (13/14).

PCA (Property Council of Australia). 2000. Unleashing Corporate Property—Getting Ahead of the Pack.

Rodgers, P.A. 2004. "Performance matters: How the high performance business unit leverages facilities management effectiveness." Journal of Facilities Management 2(4).

Then, Danny Shiem-shin. 2004. The Future of Professional Facility Management Education in the Asia-Pacific Region. New World Order in Facility Management Conference, Hong Kong.

Warren, Clive, and Sherman Heng. 2005. FM Education: Are We Meeting Industry Needs? Pacific Rim Real Estate Society Conference, Melbourne, Australia.

4

A Comprehensive Strategy for Workforce Development

The enhancement and sustainment of core competencies require organizational leadership, a comprehensive strategy for workforce development, sustained investment of resources, and a system to measure progress toward workforce development goals. A comprehensive strategy for developing the workforce integrates a range of approaches for recruiting, advancing, and retaining outstanding individuals who can acquire and use core competencies for facilities asset management. Professional enrichment approaches go beyond providing training seminars: They extend to providing opportunities for education through degree programs and online courses, mentoring, professional certification, participation in professional societies, and the acquisition of knowledge through research. Given the negative image of the government as an employer of choice (PPS, 2006), federal organizations will also need to shine a spotlight on the advantages of employment by the federal government and the opportunities it opens up if they are to recruit and retain outstanding professionals.

CREATING AN ENVIRONMENT FOR PROMOTING AND SUSTAINING CORE COMPETENCIES

Identification, development, and sustainment of core competencies for federal facilities asset management require long-term organizational leadership and commitment. Professional development must be valued by the leadership at all levels and aligned with the organization's mission and functions. It should be an institutionalized part of a talent retention and succession planning system. To make this happen, professional development will need to be assimilated into the organization's culture, integrated into the performance evaluation system, and supported through the annual budget.

The committee's research into the various approaches to professional development yielded a set of what are sometimes called "truisms" (Rose and Nicholl, 1997). These truisms, or self-evident truths, can nonetheless serve as guiding principles for promoting an organizational philosophy that embraces professional development:

- Every person can learn.
- Individuals learn at different rates in different ways.
- Learning is a lifelong process.
- Every person wants to do a good job.
- Self-esteem affects learning; learning enhances self-esteem.
- Success promotes other successes.
- Education and learning are shared responsibilities.
- People are accountable for their own decisions and actions.
- Appreciation of individuality and diversity is important; cultural diversity enhances education.
- Global awareness and understanding are essential components of education.
- Working cooperatively is essential in a competitive world.
- The education process requires innovation, risk taking, and the ability to manage change.
- Continuous improvement is desirable and possible.
- A healthy organization provides access to information.

Creating an environment that supports professional development begins with the recognition that (1) an individual does not acquire capabilities in a linear fashion and (2) the skills required of an employee change over a career. Entry-level professionals in facilities asset management divisions often have a degree in a technical field such as architecture or engineering. To advance to higher positions, individuals will need to refine their management and behavioral skills as they advance their technical skills. An embedded challenge to improving and maintaining management skills is that professionals move both laterally—that is, away from their initial discipline—and upward, into the management arena.

A professional with a honed set of behavioral and management skills will probably achieve supervisory positions, whereas one who focuses only on technical skills will probably reach a career plateau. To reach the executive level of an organization, an individual typically will need to demonstrate leadership skills and enterprise knowledge (Figure 4.1).

Individuals who progress in their careers typically are self-motivated and willing to take the initiative in expanding their knowledge through reading,

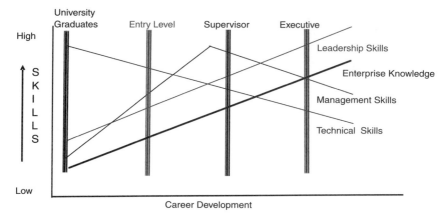

FIGURE 4.1 Changing skill sets for career progression. SOURCE: Badger and Smith (2006).

networking, attending courses, and life experience. However, organizations can provide additional support or offer incentives that benefit both the individual and the organization. One study found that "best practice organizations motivate employees as individuals and as groups to meet or exceed accepted levels of performance by establishing incentives that encourage effective decision making and reward extraordinary performance" (NRC, 2004, p. 112). Incentives that might be used by federal organizations come in many forms. Some organizations reimburse tuition, which motivates staff to enhance their skills. Tuition reimbursement could be targeted at individuals who pursue degrees in business or public administration. Similarly, tuition for executive or leadership training seminars could be reimbursed. Such programs can strengthen the skills base while motivating individuals to become more effective facilities asset managers, thereby increasing their value to the organization. Another incentive is allowing an individual time to pursue training during regular business hours as opposed to evenings and weekends. Individuals or teams who exceed performance expectations can be formally recognized. And promotions can be based, in part, on an individual's acquisition of skills that enhance the organization's core competencies.

ELEMENTS OF A COMPREHENSIVE WORKFORCE DEVELOPMENT STRATEGY

A comprehensive workforce development strategy should use a variety of approaches to recruit personnel to fill skill gaps and to enhance the skills of all staff at all stages of their careers. These approaches will need to be tailored to the types of skills being acquired. For example, gaps in technical or business-related skills can be filled by hiring recent graduates with degrees in architecture, engineering, or business or public administration and then giving them on-the-job training in the fundamentals of facilities asset management. All staff will need to update their technical skills during the course of their careers; without such renewal, their technical proficiency would decline. Business skills, on the other hand, might be strengthened by hiring experienced professionals from private companies, which value individuals who understand financial and risk management. Or, to acquire leadership skills, a facilities asset management division might hire experienced military engineering officers who have undergone leadership training. Because enterprise knowledge comes from continuous application and learning through a set of experiences that crosses many disciplines, it would be best advanced by leading employees along an in-house career path. Enterprise knowledge reflects a level of talent, experience, and critical analysis that is essential to an organization's management and leadership.

A variety of educational techniques and concepts are available for training and skills development:

- Tier-one learning: initial learning/self-learning online with computer support.
- Tier-two learning: experience-based, peer-to-peer learning in a classroom setting.
- On-the-job training.
- In-house classroom training by experts in federal subject matter.
- Technology-assisted learning or computer-aided courses.
- Training outsourced to professional consultants and specialized organizations.
- Intra-/interagency training outsourced to or conducted in collaboration with other agencies.

A comprehensive workforce development strategy should coordinate the educational materials and techniques with the type of training and the individuals being trained. For example, some knowledge can be imparted online from anywhere or at any time. Other training is best conducted in a classroom setting with face-to-face interaction among the participants.

Recruiting People with the Appropriate Skills

To overcome the negative image of the government as an employer, significant effort will be required when trying to attract recent graduates or experienced professionals to work in federal facilities asset management divisions. These recruiting efforts will have to highlight all the benefits of public service as well as the opportunities that may not be available in private firms. Examples of such benefits include the government's health care and pension plans, support for the pursuit of advanced degrees in competency-related fields, the opportunity to work at specialized facilities that have missions such as space exploration or scientific research, the chance to work or travel abroad (in the case of divisions such as the Office of Overseas Building Operations), and the potential for a stable work location (which might appeal to more mature recruits).

Perhaps most important, the job descriptions used to advertise openings and to evaluate candidates should be updated to reflect changing core competency requirements. OPM's classification and job grading standards (OPM, 2002) classify facilities professionals in the GS-1600 group, Equipment, Facilities, and Services. The GS-1600 group includes facilities personnel whose duties are to advise on, manage, or provide instructions and information on the operation, maintenance, and use of equipment, shops, buildings, laundries, printing plants, power plants, cemeteries, or other government facilities, or other work involving services provided predominantly by tradespeople or manual laborers. These oversight personnel need to have technical or managerial knowledge and ability, plus a practical knowledge of trades or manual labor operations.

Two facilities-management-related categories of employee are GS-1601, General Facilities and Equipment Series, and GS-1640, Facilities Management Series. These job classifications are broad and cover both managerial work and the supervision of administrative activities related to building operation and maintenance. Typical job titles include facility operations specialist, public works manager, or just plain "manager." These titles do not describe the broader, higher-level qualifications needed by a facilities asset manager or a senior real property officer.

This OPM documentation often serves as the informational basis for attracting talent to federal career opportunities. In addition, it provides the basis for the point system that determines a job's classification and salary level. The descriptions must be revised from time to time to reflect new realities and required competencies, so that organizations can recruit people with the necessary skills. Classification and compensation levels may also need to be revised to attract high-quality professionals. A government-wide revision of descriptions will require cooperation among facilities asset management divisions, senior real property officers, chief human capital officers, and the OPM.

In the near term, it will be difficult for facilities asset management divisions to hire recent graduates with degrees in facilities asset management, simply because few colleges and universities offer such degrees. Some universities are offering joint management and engineering programs whose graduates earn a master's degree in engineering and in business administration (NAE, 2004). Others are offering 5-year engineering programs that award both a bachelor's degree and a master's degree. In addition to hiring university graduates with facilities management and joint degrees, federal organizations will probably need to recruit crossover graduates in architecture, engineering, business, construction, public administration, or law and then train them in the fundamentals of facilities asset management. Facilities asset management divisions should always seek to recruit individuals with behaviors that suit them to work in a collaborative, innovative, dynamic environment.

Some federal organizations may need to use short-term approaches to fill critical skills gaps. Contracting with federal retirees who have essential technical knowledge of specific types of facilities or systems is one way to do this. Ideally, these contracted experts would be asked to mentor new hires and pass along their knowledge. A federal organization might want to contract with a nongovernmental organization to acquire some financial analyses and services.

STRATEGIES FOR PROFESSIONAL DEVELOPMENT

Because facilities asset management is an evolving profession, asset management divisions will need to find ways to import ideas, technologies, and research-based knowledge to support their decision making and improve their outcomes. Encouraging their employees to become involved in professional organizations and to pursue professional certifications, mentoring, and acquiring knowledge through research is a good way to do this. Such measures also help individuals to progress in their careers and become more valuable to the organization.

On-the-Job Training and Experience

An entry-level individual with a bachelor's degree in a technical area can gain an understanding of facilities asset management and learn how to apply theory to operations through on-the-job training. Developmental assignments have also been used by some federal organizations for enhancing staff skills while also improving organizational processes. The Defense Finance and Accounting Service, for example, gives staff the opportunity to work for a week in an area completely different from their own. Such assignments allow managers to "identify good people, free up their time, and use them to help solve departmental problems." In turn, employees "felt good about being able to learn new skills and also to contribute to solving problems within their work organizations" (Mathys and Thompson, 2006, p. 23).

A rotational program for executives allows managers to rotate among divisions to allow them to gain experience, enterprise knowledge, and behavioral and leadership skills. Similarly, assigning a facilities asset manager as a deputy to a division executive might be an excellent way for that manager to gain enterprise knowledge quickly.

Involvement in Professional Organizations

One important component of professional development is joining a professional society, becoming active in one of its peer groups, and attending professional development training along with other senior and intermediate leaders. Such participation would ensure that an "idea transfer model" is in place—that is, that facilities asset management professionals will learn and then transfer new ideas, technologies, and/or research results back to the facilities asset management division and the larger organization. To satisfy the demands of the private sector, a number of professional organizations have emerged whose members possess a broad array of cross-disciplinary backgrounds, skill sets, management methodologies, and leadership approaches. Federal organizations would do well to take advantage of membership in such organizations to further the professional development of their own workforces.

As an example, the Construction Industry Institute (CII) serves as a forum in which managers and executives can identify and collaborate on new ideas, which they take back to their own organizations. CII's Education Committee has come up with an array of training deliverables (Figure 4.2), including technology-assisted learning, that it hopes to market to professionals who might do well to learn from the relevant experience of CII.

Starting at the top and going clockwise, professionals may participate in research project teams, learn from prior CII research, network with other team members, problem-solve, and join ongoing teams. Research summaries and documents for reading and studying are available to individuals. The leadership of the 90 companies and organizations that are members of CII recognizes the

FIGURE 4.2 Training modules offered by CII to share its experience. SOURCE: CII (2007).

educational and professional growth of their respective representatives within the different components of the learning wheel. Professionals who actively engage in research efforts appear to derive the greatest benefit.

Other organizations have developed targeted training. For example, APPA has come up with a 4-week leadership course for educational facilities managers. The course has four tracks—individual, interpersonal, managerial, and organizational effectiveness—each of which has a different perspective and is devoted to a different leadership skill (APPA, 2007). Federal managers will need to determine which types of training are best suited to delivering essential skills to their organizations.

Certificate Programs

Federal executives should also consider using the certification programs of professional organizations to improve the competencies and skills of their facilities asset managers. The International Facility Management Association has a well-established Certified Facility Manager program. The certification procedure assesses competency in the field based on work experience and education and the ability to pass a comprehensive examination. The examination tests for core skills and knowledge in planning and project management; real estate; leadership and management; finance; technology; operations and maintenance; quality assessment and innovation; human and environmental factors; and communication (IFMA, 2007).

To certify their technical skills, engineering graduates must take an engineer-in-training exam (now called the Fundamentals of Engineering exam) and document 5 years of engineering design experience before becoming a registered Professional Engineer (PE). Each of the 50 states issues a PE license, and many require continuing education courses to maintain the license. Most states have established a code of ethics for their PEs.

The American Institute of Contractors has devised an exam and set experience criteria for becoming a Certified Professional Constructor. The Project Management Institute has created a certification procedure for project and program managers. The Construction Management Association of America has a construction manager certification program with a component specifically for federal employees.

As described in Chapter 3, the Defense Acquisition University (DAU) has a training course for federal facilities asset managers, including a module for senior managers. Facilities asset management divisions could form an interagency alliance to develop a government-wide certification program for facilities asset managers using the DAU model as a starting point.

Mentoring

Mentoring is a relationship in which one person (the mentor) invests time, knowledge, and effort in enhancing the growth, knowledge, and skills of another person (the protégé). Mentoring can occur informally, when two people meet and end up in a work-related relationship, or through formal, structured programs. Formal mentoring programs in federal organizations can be a good means for transferring technical, business, and leadership skills and enterprise knowledge from senior employees to new recruits and should be considered as part of an overall workforce development strategy.

A formal mentoring program is designed to achieve specific objectives related to developing and sustaining core competencies—for instance, the transfer of knowledge about facility design, construction, or operations. Mentors are matched with protégés, and each is given a set of well-defined expectations. Mentors are typically expected to invest time, knowledge, and effort; provide positive encouragement, feedback, and reinforcement to the protégé; and identify activities and resources that the protégé can use to increase his or her knowledge of the specified subject area and professional contacts. The protégé is expected to be eager to learn; willing to tackle difficult problems and to persevere in the face of obstacles; work as a team player; and enhance his or her knowledge of the area of expertise. In doing so, the protégé can enhance his or her professional development and value to the organization. A mentoring program can also be used to enhance the skills base of an individual. By mentoring mid-career and entry-level staff, soon-to-retire senior employees can transfer institutional knowledge about critical systems and processes.

Knowledge Acquisition Through Research

Research is the cutting edge of any discipline. Research-based knowledge can serve professionals and also those they serve. For the evolving profession of facilities asset management, knowledge development must play a significant, extensive role. Knowledge can be gained through a process that can provide a continuous improvement environment for both organizations and individuals. One process by which knowledge can be gained is shown schematically in Figure 4.3.

The most immediate, applicable form of knowledge is that which is gained on the job. Leaders of facilities asset management divisions need to be aware of the knowledge that resides in their workforce and aware of where there are knowledge gaps. For example, does the workforce have the capacity to create a facilities asset management plan that incorporates acknowledged best practices? If not, what research-based knowledge is needed to improve those capabilities?

If a facilities asset management division does not have adequate in-house resources to conduct research that will improve the knowledge base, it can engage

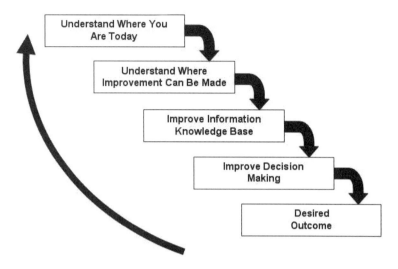

FIGURE 4.3 Example of a knowledge development process.

an outside source of expertise from a university or a professional society or hire a private consultant for this purpose. Relying on outside expertise to conduct research can provide the independent voice needed for credibility. When research is undertaken federal personnel should be involved in guiding the effort to ensure that it is applicable to the federal operating environment. Their involvement will also benefit the staff's experiential growth and will increase the chances that the results of a study will be implemented. Involvement often enhances buy-in, promotes implementation, and facilitates culture changes.

Investment of Dedicated Resources

A long-term organizational commitment is needed to create an environment that supports professional development and to optimize the investment of staff time and monetary resources. Private companies recognize that training is necessary to maintain an effective executive workforce, and they invest accordingly. Some examples of budget allocations for professional development follow:

- Motorola University budgets 3.6 percent of its payroll on training, some $120 million a year.
- Saturn Corporation, with its emphasis on shared values, cooperative decision making, and teamwork, expects every employee to undertake 100

hours a year of formal education. General Electric places 13,000 employees in a 2-day course in thinking skills and problem-solving.
- Fel Pro reimburses employees for tuition costs and pays bonuses to people who earn degrees.
- The Arup Partnership allows 10 percent of all employee time to be devoted to continuing education.

Executives of the federal government's Defense Finance and Accounting Service (DFAS) identified the core competencies its workforce needs to achieve the organization's mission. To instill these competencies DFAS is committed to spending 5 percent of its labor budget on training. In 2003, the agency spent about $1,800 per individual for training, or about 4 percent of its total payroll (Mathys and Thompson, 2006).

Based on these examples and the committee's collective experience, a comprehensive strategy for imbuing an organization's workforce with the required competencies would cost 2 to 5 percent of the organization's annual personnel budget.

PERFORMANCE MEASUREMENT

One critical element of a workforce development strategy is the establishment of a system to measure progress in developing and sustaining core competencies. A system for monitoring performance can serve as a basis for continuous improvement. Indeed, the monitoring process itself often results in better performance. The old adage "what gets measured, gets improved" or, better yet, "what gets measured and reported, gets improved quickly" makes it important to measure the correct things.

Organizational performance is tied to mission accomplishment, proper stewardship of the assets, and effective workforce management. Ongoing self-assessment is key for an organization to have an accurate picture of how it is performing. This assessment should ask "Where are we going?" and "Can we get there from here?" The second of these questions is usually where the challenge lies. Answering it requires an in-depth review of skill sets, priorities, and mission alignment. Gaps in the required skill sets could come from insufficient training of the workforce or from the lack of suitable personnel. It is critical to assess priorities, which will probably change over time to reflect new demands on the organization. Federal organizations need to measure the following:

- Performance in meeting specific targets, which will vary from organization to organization.
- Performance in complying with the strategic plan of the facilities asset management division. Some key performance indicators are customer satisfaction, the maintenance backlog, and financial measures.

- Progress in developing the required core competencies.
- Progress in attracting, hiring, and retaining highly competent professionals.

The Government Performance and Results Act of 1993 requires federal organizations to establish performance measurement systems for assessing the outcomes of their programs. In the 14 years since this legislation was enacted, performance measurement has become an established process within the federal government.

Several performance measurement systems are well known, including the Malcolm Baldrige National Quality Program, the Balanced Scorecard (BSC), and the Strategic Assessment Model (SAM). SAM was designed by APPA specifically for facilities management divisions and incorporates elements of the Baldrige program and others of the BSC. Because the BSC is a well-established system that federal organizations are already using, the committee recommends its continued use for measuring progress in developing and sustaining core competencies for federal facilities asset management divisions. The Balanced Scorecard (BSC) is widely used by business and government even though the goals of business and government are different (Table 4.1).

First described in 1990, the BSC is a conceptual framework for evaluating organizational performance. Organizational goals and strategies are translated into objectives, measures, targets, and initiatives. The BSC concept has evolved over time, but the four categories of performance have remained constant: financial outcomes, internal business processes, customer relationships, and learning and

TABLE 4.1 Comparing Balanced Scorecards in the Private and Public Sectors

Feature	Private Sector	Public Sector
Focus	Shareholder value	Mission effectiveness
Financial goals	Profit, market share growth, innovation, creativity	Cost reduction, efficiency, accountability to public
Efficiency concerns of clients	No	Yes
Desired outcome	Customer satisfaction	Stakeholder satisfaction
Stakeholders	Stockholders, bondholders	Taxpayers, legislators, inspectors
Who defines budget priorities	Customer demand	Leadership, legislators, funding agencies
Key success factors	Uniqueness, advanced technology	Sameness, economies of scale, standardized technology

SOURCE: Mathys and Thompson (2006).

growth. The learning and growth category focuses on the organization's workforce, one resource which enables the achievement of the organization's goals. "Balanced" refers to several qualities of the scorecard. First, there is balance across the four categories to avoid overemphasizing financial outcomes. Second, it requires both the quantitative and qualitative measurement of outcomes. Third, there is balance in the levels of analysis—from individual and work unit results to organizational outcomes (Heerwagen, 2002; NRC, 2004).

Within the four categories of performance, there is an implied alignment from performance-driven staff and an organizational culture reflected in learning and growth to operational efficiency and improvements in customer satisfaction, both of which improve financial outcomes (Mathys and Thompson, 2006). In the private sector, this alignment leads to greater stakeholder value; in the public sector, it leads to mission effectiveness.

By design, the BSC provides for cascading goals, objectives, and performance measures, from organizational missions to work units to individuals (FFC, 2004). Performance measures are the link between strategies and operations. Thus a BSC can be used to measure both the performance of a facilities asset management division in supporting the mission of its parent organization and the performance of an individual manager of facilities assets.

Figure 4.4 illustrates how the DFAS uses the BSC to link strategy and measures to organizational vision. As shown in Figure 4.4, learning and growth objectives might include enhancing employee competencies; increasing employee satisfaction; developing an environment for learning; and enhancing the organization's ability to recruit and retain employees with the required skills. The indicators of progress in achieving these objectives would appear to be a natural outgrowth of the skills gap analysis. Indicators of progress in closing the gaps could be based on the number of new hires having particular skill sets or the number of vacancies filled. Gaps in the required skill sets could also be closed through professional development and training. Indicators such as the number of hours spent in training to strengthen organizational core competencies, hours devoted to the pursuit of advanced degrees in finance and other areas, or the number of facilities asset managers who received a job-related certification all could be useful for measuring progress in this area. If training and recruitment efforts are effective, they should result in improved internal processes, more customer satisfaction, and greater cost efficiency.

Additional indicators for workforce development can be derived from guidelines of the OPM as part of its Human Capital Assessment and Accountability Framework (HCAAF). The HCAAF resource center includes guidelines for assessing strategic alignment, leadership, knowledge management, and talent management (OPM, 2007).

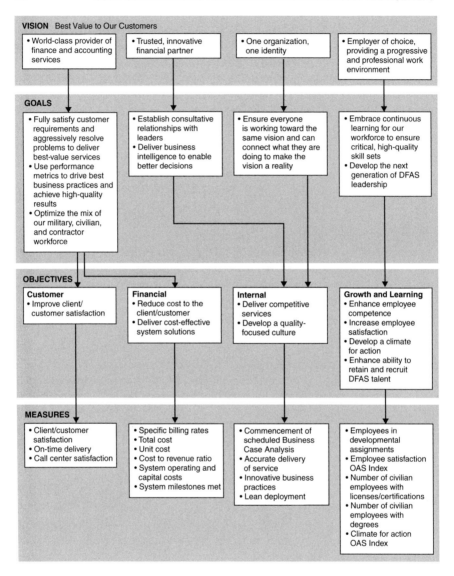

FIGURE 4.4 Example of the BSC from the Defense Finance and Accounting Service. SOURCE: Mathys and Thompson (2006).

REFERENCES

Badger, W., and J. Smith. 2006. Great Leadership Skills and Traits: The Faculty's Secret Enabler. Proceedings of the 2nd Specialty Conference on Leadership and Management in Construction, Grand Bahamas, May 4-6.

CII (Construction Industry Institute). 2007. Implementing CII Practices—The Implementation Planning Model: Steps to Success. Austin, Tex.: CII.

Heerwagen, J.H. 2002. "A balanced scorecard approach to post-occupancy evaluation: Using the tools of business." In Learning from Our Buildings: A State of the Practice Summary of Post-Occupancy Evaluation. Washington, D.C.: National Academy Press.

Horey, J. 2004. Technical Report 1148, Competency Based Future Leadership Requirements. Arlington, Va.: U.S. Army Research Institute for the Behavioral and Social Sciences.

IFMA (International Facility Management Association). 2007. Certified Facility Manager Credentials. Accessed on August 13, 2007, at <www.ifma.org/learning/fm_credentials/forms/cfmbrochure.pdf>.

Mathys, N.J., and K.R. Thompson. 2006. Using the Balanced Scorecard: Lessons Learned from the U.S. Postal Service and the Defense Finance and Accounting Service. IBM Center for the Business of Government. Available online at <www.businessofgovernment.org>.

NAE (National Academy of Engineering). 2004. The Engineer of 2020: Visions of Engineering in the New Century. Washington, D.C.: The National Academies Press.

NRC. 2004. Investments in Federal Facilities: Asset Management Strategies for the 21st Century. Washington, D.C.: The National Academies Press.

OPM (Office of Personnel Management). 2002. Handbook of Occupational Groups and Families. Classification and Job Grading Standards. Washington, D.C.: OPM.

OPM. 2007. Human Capital Assessment and Accountability Framework Resource Center. Available at <//www.opm.gov/hcaaf_resource_center/>. Accessed May 17, 2007.

PPS (Partnership for Public Service). 2006. Back to School: Rethinking Federal Recruiting on College Campuses. Washington, D.C.: Partnership for Public Service.

Rose, C., and M.J. Nicholl. 1997. Accelerated Learning for the 21st Century: The Six-Step Plan to Unlock Your Master-Mind. New York, N.Y.: Delacorte Press.

5

Core Competencies for Federal Facilities Asset Management: Findings and Recommendations

Chapters 1 through 4 provide context and address the statement of task by identifying forces affecting the federal government, the core competencies needed to manage facilities assets, development strategies, and indicators for measuring progress toward the development of core competencies. Chapter 5 extracts the key findings from the study and presents seven recommendations for federal facilities asset management.

FINDINGS

Finding 1: Previous NRC reports on the management of federal facilities recommended that federal organizations approach facilities asset management with the mindset of an owner; align their facilities portfolios to their organization's missions through strategic decision making; take a life-cycle management approach; and measure performance to continuously improve facilities management.

The report *Stewardship of Federal Facilities: A Proactive Strategy for Managing the Nation's Public Assets* (NRC, 1998, p. v) addresses the issue of ownership as follows:

> The ownership of real property entails an investment in the present and a commitment to the future. Ownership of facilities by the federal government, or any other entity, represents an obligation that requires not only money to carry out that ownership responsibly, but also the vision, resolve, experience, and expertise to ensure that resources are allocated effectively to sustain that investment. Recognition and acceptance of this obligation is the essence of stewardship.

On page 63, the report identified the four most important elements in creating a climate for stewardship:

- Leadership by agency senior managers.
- The establishment and implementation of a stewardship ethic by facilities program managers and staff as their basic business strategy.
- Senior managers and program managers who create incentives for successful and innovative facility management.
- Agency strategic plans that give suitable weight to effective facilities management.

Outsourcing Management Functions for the Acquisition of Federal Facilities states that "when acquiring facilities, a federal agency assumes an ownership responsibility as a steward of the public's investment" (NRC, 2000, p. 46). It notes that ownership and management functions are not the same. Owners establish objectives and make decisions. They determine the need for facilities, develop project scopes, balance conflicting priorities, establish parameters (for cost and duration, for example), and determine positions in negotiations. Managers perform ministerial functions that implement the owner's decisions and oversee accomplishment of the task. Ownership and management functions are equally important and must be balanced if organizations are to effectively achieve their missions. A smart owner of facilities has the skill base—usually a staff with the professional qualifications and authority—necessary to plan, guide, and evaluate the facility acquisition process. A smart owner focuses on the "relationship of a specific facility to the successful accomplishment of an organization's business or overall mission" (NRC, 2000, p. 50).

Investments in Federal Facilities: Asset Management Strategies for the 21st Century (NRC, 2004) recommends that federal organizations integrate facilities considerations into their strategic planning to provide decision makers with better information about the total long-term costs and consequences of a particular course of action. It also recommended that the senior facilities asset manager participate in the organization's strategic planning at the executive level so that he or she can translate between its missions and its facilities portfolio and clearly communicate how real estate and facilities can support the missions.

The 1998 and 2004 reports both recommend a life-cycle approach for managing facilities. All three reports recommend the continuous measurement of performance to improve federal facilities asset management. Performance measures help organizations to understand why particular decision-making processes or management practices succeeded or failed and where objectives are not being met or where they are being exceeded. Managers can then "investigate the factors or reasons underlying the performance and make appropriate adjustments" (NRC, 2004, p. 65).

Finding 2: Geopolitical and socioeconomic trends, rapid advances in technology, reliance on outside contractors, budget pressures, heightened focus on sustainable development, and government-wide reforms require a paradigm shift in both federal facilities management and the core competencies of facilities asset management divisions.

The high risk associated with federal facilities management has been recognized by the Government Accountability Office (GAO) and the President's Management Agenda (PMA). The poor condition of some federal facilities places people's health, safety, and welfare at risk, hinders the accomplishment of missions, and becomes a significant long-term financial burden. Because the federal government is the single largest owner of facilities in the United States and, indeed, in the world, it can play an important role in sustainable development by reducing the environmental impacts of facilities.

Current geopolitical conditions indicate that the nation's focus on homeland security and global terrorism will continue. Information technology is changing not only the methods by which the physical infrastructure is managed but also the infrastructure itself. Changes in government paradigms, the growing national fiscal imbalance, and changes in the workforce all have tangible implications for federal facilities asset management. Unless facilities management policies and practices change significantly, the factors that make federal facilities a high-risk proposition—lack of alignment with missions, deteriorating condition, and obsolescence—will only worsen.

Finding 3: Significant reductions in the federal workforce through across-the-board cuts and hiring freezes have resulted in a workforce whose skills are not aligned with new technologies or business practices. These deficiencies will be exacerbated within the next few years as the most experienced facilities managers—the baby boom generation—retires.

In the wake of the downsizing efforts of the 1990s, the federal government is now challenged to acquire and develop staff whose numbers, skills, and deployment meet organizational needs (GAO, 2001). The PMA notes that without proper planning, the skill mix of the federal workforce will not be suitable for changing missions.

Within the entire executive branch, 42 percent of the Senior Executive Service, which includes most managerial and policy positions, is projected to retire by 2010. One potentially enduring outcome of these retirements is the loss of the institutional knowledge base built up over decades, including knowledge of the design, construction, and operation of specialized facilities, particularly for military and complex civil works projects. At a minimum, loss of this knowledge could result in project-specific inefficiencies; in the worst case, it could negatively impact military preparedness and jeopardize national security.

Finding 4: Efforts to recruit recent graduates and experienced professionals to federal service are hampered by the poor image of the federal government as an employer, cumbersome hiring processes, and outdated job descriptions. The GS-1600 series, which is the basis for hiring and compensating facilities managers, is based on an old paradigm and does not reflect new realities or required core competencies.

As baby boomers retire, the federal government will be challenged to attract a new generation of workers into federal service—a generation that is technology-savvy and that was born in the Information Age. This generation will be accustomed to connectivity, mobility, and the availability of information. A recent survey found that many college students were unaware of federal career opportunities, and others said that "too much bureaucracy" and "low salaries" were reasons not to work for the government (PPS, 2006).

Another barrier to recruitment is the GS-1600 job classification series, which is the basis for recruiting new employees and talent to the federal facilities asset management arena and which influences compensation levels. The current descriptions focus on overseeing the operation, maintenance, and use of facilities and equipment and do not take into account the broader, higher-level qualifications required for facilities asset managers.

Finding 5: Of all the stakeholders involved in funding, programming, designing, constructing, operating, and maintaining federal facilities, facilities asset management divisions are the only ones involved in all phases. To effectively support their organization's missions, facilities asset managers must integrate people, processes, technologies, services, and knowledge.

Within the federal government, facilities investment and management decisions involve multiple stakeholders (Congress, the executive branch, the public), decision makers (Congress, the Office of Management and Budget, senior executives of departments and agencies), and more than 30 facilities asset management divisions.

Facilities asset management divisions perform a wide range of functions using different technologies and involving different operational units (construction, design, energy, engineering, electrical, environmental, fire, forestry, historical, information technology, janitorial, landscaping, maintenance, mechanical, parking, power, safety, sanitary, security, space, traffic, and others). This diversity of functions means that facilities asset management divisions must serve as "connected integrators" of diverse stakeholders, functions, and services across the life cycles of facilities.

Finding 6: The federal operating environment is dynamic and requires that facilities asset management divisions be able to innovate to address changing functional requirements and to take advantage of opportunities

for improvement as they arise. Leadership skills—the ability to influence beyond one's authority—are essential.

The GAO and others have recognized the importance of leadership skills for managing the change that is necessary for transforming federal programs and practices. Leadership requires a wide-ranging balance of the business, technical, management, and behavioral skills and values that allow an individual to make effective decisions, influence others, and lead them to success.

Finding 7: To ensure that core competencies are established and sustained, federal organizations need a comprehensive workforce development strategy. They will also need to provide a long-term commitment to and investment in the professional development of facilities asset managers.

A comprehensive strategy for workforce development requires a set of approaches for attracting, advancing, organizing, motivating, and retaining outstanding individuals. Long-term professional development of staff will need to go beyond training seminars. It should provide opportunities for education through degree programs, online courses, mentoring, and professional certification and should encourage participation in professional societies. The creation and implementation of such a strategy requires an environment that encourages learning and the development of core competencies. Fostering such an environment will in turn require organizational leadership and commitment, adoption of core values, and the sustained investment of resources.

Finding 8: Where information gaps exist within a facilities asset management division, facilities asset managers will need to find ways to import research-based and experience-based knowledge.

Because facilities asset management is evolving as a profession, facilities asset management divisions will need to find ways to import ideas, technologies, and research-based knowledge to support their decision making and improve their outcomes. Leaders of facilities asset management divisions need to understand the level of knowledge that resides in their workforces and where there are gaps in that knowledge.

As facilities asset management in the private sector evolves as well, many new ideas and principles have emerged from that sector that can help federal facilities asset managers understand where improvements can be made. Federal organizations can encourage individuals to be involved in professional organizations and to seek professional certification, and they can establish processes for the transfer of knowledge from outside sources into federal facilities asset management divisions.

Finding 9: A system for measuring progress in developing and sustaining core competencies is a critical element of a comprehensive workforce development strategy.

The systematic measurement of performance can bring about continuous improvement. Indeed, the very process of measurement often improves performance. Organizational performance is tied to mission accomplishment, proper stewardship of assets, and effective workforce management. Ongoing self-assessment is required if an organization is to have a good picture of where it stands in achieving its goals. Such an assessment should ask "Where are we going?" and "Can we get there from here?"

REQUIRED CORE COMPETENCIES

Based on its literature review, briefings, interviews, current geopolitical and socioeconomic trends, and the experience and knowledge of its members, the committee concluded that three essential areas of expertise are required by federal facilities asset management divisions through 2020 and beyond:

- *Integrating* people, processes, places, and technologies by using a life-cycle approach to facilities asset management;
- *Aligning* the facilities portfolio with the organization's missions and available resources;
- *Innovating* across traditional functional lines and processes to address changing requirements and opportunities.

The required skills base includes a balance of technical, business, and behavioral capabilities and enterprise knowledge. Technical capabilities in fields such as engineering and architecture are essential for life-cycle facilities management. Technical capabilities include knowledge of design and construction; facilities-related systems and their operations and maintenance; acquisition and project management processes; regulations and procedures; information technology and building technology; and analytical skills. Business capabilities include strategic planning and resource management to support an organization's missions. Behavioral capabilities involve the leadership, communication, negotiation, and change management skills required to integrate functions, people, and processes across traditional lines and the capacity to innovate within a dynamic operating environment. Enterprise knowledge includes an understanding of the facilities portfolio and how to align it with the organization's missions; an understanding of the organization's culture, policy framework, and financial constraints; agency inter- and intradependencies; and the workforce's capabilities and skills. Facilities asset managers[1] will require technical skills, enterprise knowledge, behaviors, and

---

[1] An individual in a facilities asset management division whose primary responsibilities involve some

other personal characteristics that allow them to work in a team-oriented environment and to support achievement of a life-cycle approach for facilities asset management. Together, the three essential areas of expertise and the skills base make up the core competencies for federal facilities asset management divisions through 2020 and beyond. The committee's recommendations for developing and sustaining core competencies follow.

RECOMMENDATIONS

Recommendation 1: To effectively manage federal facilities portfolios through 2020 and beyond, federal organizations and their facilities asset management divisions should operate within the overall framework depicted in Figure 5.1.

Recommendation 2: To develop its core competencies, each federal facilities asset management division should first identify the functions and skills it will need to perform or oversee in support of its organization's missions. Although similarities will emerge, the unique aspects of each organization will become apparent only after a careful analysis of current and future functional requirements.

Recommendation 3: Each federal facilities asset management division should also conduct an analysis that compares its current skills base to the skills base required for facilities asset management in 2020. The analysis should account for planned or potential changes in missions, requirements, and technologies through 2020 and should identify actions needed to close any skills gaps.

Recommendation 4: Federal organizations should develop a comprehensive, long-term strategy to acquire, develop, and sustain a workforce with the required core competencies for facilities asset management. Senior executives should show their commitment and provide the resources necessary for individuals to develop and refine their skills, including leadership skills, through a continuum of experience and opportunities.

Recommendation 5: To overcome barriers to recruiting and hiring individuals with required skills and capabilities, the directors of federal facilities asset management divisions and senior real property officers should collaborate with the Chief Human Capital Officers Council and

---

aspect of managing the organization's portfolio of facilities. Such individuals include facilities program managers, planners, architects, engineers, and project, operations, or maintenance managers.

FINDINGS AND RECOMMENDATIONS

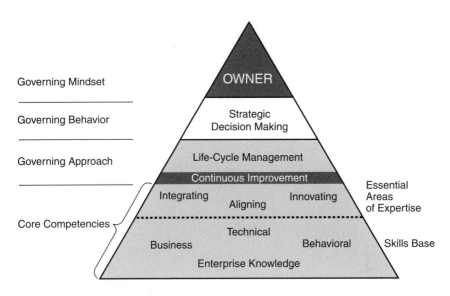

FIGURE 5.1 Recommended framework for effective federal facilities asset management.

the Office of Personnel Management to revise the GS-1600 job classification series.

Recommendation 6: Federal facilities asset managers should seek to expand the knowledge base related to facilities asset management and use the results to improve decision making and achieve the desired outcomes. Knowledge can be transferred through involvement in professional societies, certification programs, and research using in-house or outside expertise.

Recommendation 7: Federal organizations should use a Balanced Scorecard approach for measuring progress in developing and sustaining core competencies for facilities asset management through 2020 and beyond.

REFERENCES

NRC (National Research Council). 1998. Stewardship of Federal Facilities: A Proactive Strategy for Managing the Nation's Public Assets. Washington, D.C.: National Academy Press.

NRC. 2000. Outsourcing Management Functions for the Acquisition of Federal Facilities. Washington, D.C.: National Academy Press.

NRC. 2004. Investments in Federal Facilities: Asset Management Strategies for the 21st Century. Washington, D.C.: The National Academies Press.
PPS (Partnership for Public Service). 2006. Back to School: Rethinking Federal Recruiting on College Campuses. Washington, D.C.: Partnership for Public Service.

Appendixes

Appendix A

Biographies of Committee Members

RADM David J. Nash, *Chair*, USN Civil Engineering Corps (retired), was elected to the National Academy of Engineering in 2007 for leadership in the reconstruction of devastated areas after conflicts and natural disasters. He has established a consultancy, Dave Nash and Associates, and is also the president of BE&K Government Group, a wholly owned subsidiary of BE&K, Inc. Headquartered in Birmingham, Alabama, BE&K is a privately held international design-build firm that provides engineering, construction, and maintenance services for process-oriented industries, commercial real estate projects, and the U.S. federal government. In 2003 and 2004, RADM Nash served as the director of the Iraq Reconstruction Program. He was formerly president of PB Buildings and manager of the Automotive Division of Parsons Brinckerhoff Construction Services, Inc. RADM Nash completed his 33-year career in the U.S. Navy as the chief of the Naval Facilities Engineering Command and chief of civil engineers. He previously served as vice chair of the NRC Committee on Business Strategies for Public Capital Investment, which produced the study *Investments in Federal Facilities: Asset Management Strategies for the 21st Century.* He is a member of the National Academy of Construction, the Society of American Military Engineers, the American Society of Civil Engineers, the American Public Works Association, the National Society of Professional Engineers, the Institute of Electrical and Electronics Engineers, and the American Society of Quality Control.

William W. Badger is the director of and a professor at the Del E. Webb School of Construction at Arizona State University. Dr. Badger has led and managed the school, which is in the College of Engineering and Applied Sciences, since 1987. Before joining the school, Dr. Badger had a distinguished 26-year career in the

U.S. Army Corps of Engineers, serving in China, Viet Nam, Saudi Arabia, Europe, and the United States. Before leaving the Army he was colonel and chief engineer for the United States Military Academy at West Point, from 1983 to 1985. At the academy he was responsible for all planning, engineering, construction, and maintenance. His current research interests include construction management, contracting, construction financial management and cost control, computer applications in construction, and facility operation and management. Dr. Badger holds a B.S. in mechanical engineering from Auburn University, an M.S. in civil engineering from Oklahoma State University, and a Ph.D. in soil mechanics from Iowa State University.

Jennifer J. Compagni is a human resources consultant assisting clients to develop and implement HR processes, including compensation programming and performance, position competency models, management structures, communications plans, and leadership development. Prior to establishing her consultancy she was vice president for human resources at Revlon, Inc., where her work included leadership development and team building. She also held senior management positions at Warner-Lambert/Pfizer and ADAMS USA. Ms. Compagni holds a B.A. in economics from Siena College and a master's in industrial and labor relations from Cornell University.

Dennis D. Dunne is president of dddunne & associates, a consulting firm headquartered in Scottsdale, Arizona. Mr. Dunne is the former chief deputy director for the California Department of General Services. His 25-year career was spent creating, organizing, directing, and consulting with large public sector departments centered on the built environment. The thrust of these efforts was the development of a customer-driven, continuously improving, life-cycle-oriented culture. As chair of the Policy Executive Committee, he helped develop the Excellence in Public Buildings Initiative, which expresses California's commitment to a set of policies, guidelines, procedures, and practices that will lead to sustained excellence in the planning, design, construction, operation, and performance evaluation of public buildings. Mr. Dunne has also been a consultant to government agencies and construction and architectural firms and has facilitated strategic planning, project management development, and construction partnerships. He was also the General Services Agency director for Santa Clara County, California, and vice president for support services for Kitchell CEM. Mr. Dunne is a member of the Board on Infrastructure and the Constructed Environment of the National Research Council.

Martin A. Fischer is an associate professor of civil and environmental engineering and the director of the Center for Integrated Facility Engineering at Stanford University. His research is in the area of construction management tools, with specific interest in the formal, computer-interpretable representation of construction knowledge and design. Current research topics are model-based constructability analysis; linking design and construction with construction method models; product, process, and organization prototyping for concurrent engineering; collaborative four-dimensional, computer-aided design; and integrated management of construction and facility information. He has a diploma in civil engineering from the Swiss Federal Institute of Technology and an M.S. in industrial engineering and a Ph.D. in civil engineering from Stanford University.

Michael J. Garvin, Ph.D., P.E., is an assistant professor in the recently established Myers-Lawson School of Construction at Virginia Tech. His research and education pursuits are geared to fundamentally changing how institutional owners, such as the Department of Transportation and universities and federal agencies, make constructed (or real) asset investment and financing decisions. His current research projects are developing decision support systems for portfolio-level facilities investment and financing decisions and identifying best practices for public-private partnership arrangements through case-based research. Dr. Garvin is a 2004 recipient of the Presidential Early Career Award for Scientists and Engineers, which is the highest honor bestowed by the U.S. government on outstanding scientists and engineers beginning their independent careers. He is also currently a member of ASCE's Construction Research Council and its Infrastructure Systems Committee, is on the editorial board of the journal *Public Works Management and Policy*, and is a specialty editor for the case studies division of the ASCE *Journal of Construction Engineering and Management*. He has authored or coauthored more than 30 journal articles, conference papers, and book chapters.

Alex K. Lam is vice president for global learning at CoreNet Global, a worldwide nonprofit association with a mission to advance the effectiveness of its members in delivering value to their corporations through the strategic management of corporate real estate and workplace resources. Mr. Lam conducts the Master of Corporate Real Estate (MCR) professional designation seminars in the Asia-Pacific region for CoreNet Global. He is also the founder of and the chief international director at The Workplace Institute, a think tank for workplace thought leaders. Prior to becoming a trainer/consultant, he served 23 years as general manager of facilities at Bell Canada, where he led a team of 379 professionals and was responsible for an annual operating budget of $70 million. Mr. Lam holds a B.Arch. from McGill University and a master's degree in theology from Ontario Theological Seminary.

Karlene H. Roberts is a professor in the Haas School of Business at the University of California, Berkeley, and a research psychologist at Berkeley's Institute of Industrial Relations. Dr. Roberts has expertise in the design and management of organizations and systems of organizations in which errors can have catastrophic consequences. The results of her research have been applied to programs in numerous organizations, including the U.S. Navy and the Coast Guard, the Federal Aviation Administration's Air Traffic Control System, NASA, and the medical industry. Dr. Roberts has published on a wide range of organizational risk management issues. She is a fellow of the American Psychological Association and the American Psychological Society. She has a B.A. in psychology from Stanford University and a Ph.D. in psychology from the University of California, Berkeley.

David H. Rosenbloom is Distinguished Professor of Public Administration at American University. Dr. Rosenbloom has also taught at the University of Kansas, Tel Aviv University, the University of Vermont, and Syracuse University's Maxwell School. He has more than 150 published works focusing on public administration, law, administrative theory and history, bureaucratic politics, and public personnel issues. In 1992, he was appointed to the Clinton-Gore Presidential Transition team with responsibilities for federal personnel policy and the Office of Personnel Management. In 1969, he was an American Society for Public Administration fellow at the U.S. Civil Service Commission. He is a member of the National Academy of Public Administration and the recipient of numerous awards. He was editor in chief of the *Public Administration Review* from 1991 to 1996 and currently serves on the editorial boards of about a dozen leading public administration journals. Dr. Rosenbloom received a B.A. from Marietta College and an M.A. and a Ph.D. in political science from the University of Chicago. He also received an honorary doctor of law degree from Marietta College.

Richard L. Tucker is the Joe C. Walter, Jr., Chair in Engineering (emeritus) at the University of Texas at Austin. He currently serves on the board of directors for Hill and Wilkinson, Inc., Integrated Electrical Services. He is a member of the National Academy of Engineering, a fellow of the American Society of Civil Engineers, and a member of numerous professional societies and associations, including the National Society of Professional Engineers, the American Association of Cost Engineers, and the American Society for Testing and Materials. Dr. Tucker's awards and honors include the Construction Engineering Educator Award from the National Society of Professional Engineers; the Ronald Reagan Award for Individual Initiative, Construction Industry Institute, 1994; the Michael Scott Endowed Research Fellow, Institute for Constructive Capitalism, 1990; and the Carroll H. Dunn Award, Construction Industry Institute, 1997. He has published numerous items over four decades and recently wrote "Communicating in Construction: The Path to Project Success," Chapter 1 of the 1996 Wiley Construction Law Update. Dr. Tucker has a B.S., an M.S., and a Ph.D. in civil engineering from the University of Texas at Austin.

James P. Whittaker is president of Facility Engineering Associates, P.C., where he specializes in asset management and facilities management technologies.

He is an adjunct professor in the George Mason University certificate program in facility management and has presented courses for the International Facility Management Association and the Association of Higher Education Facilities Officers (APPA) on facility management technologies and the definition of facility management core competencies. He serves on the advisory board of Brigham Young University's facility management degree program. His consulting services include evaluation of the effectiveness of facility management organizations and resource analysis for government and industry. He is a frequent contributor to the "Asset Management" column in APPA's *Facility Manager*. Mr. Whittaker holds a B.S. in civil engineering from the University of Vermont and a master's in civil engineering from the University of Colorado.

Norbert W. Young, Jr., is president of the McGraw-Hill Construction Information Group, where he is responsible for building relationships with owners, key design firms, and construction firms and for partnering with the product development, technology, sales, marketing, and production functions. Before joining McGraw-Hill, Mr. Young spent 8 years with the Bovis Construction Group, where he was president of Bovis Management Systems, which provided construction and project management services for both private and public sector clients. He was instrumental in creating an integrated approach to delivering preconstruction services, which became the standard for Bovis's operations in the United States. He started his career as an architect and gained 12 years' experience in a wide range of building types and projects. He is a member of the Urban Land Institute, the American Institute of Architects, the International Alliance for Interoperability, and the International Development Research Council. He has lectured nationally on such topics as project delivery approaches, managing the risk of the design and construction process, and outsourcing trends. Mr. Young holds a bachelor of arts degree from Bowdoin College and a master's degree in architecture from the University of Pennsylvania.

Appendix B

Committee Interviews and Briefings

2005

June 6 First Committee Meeting; Briefings by Get Moy, Office of Deputy Under Secretary of Defense–Installations and Environment; Gene Hubbard, Director of Facilities Engineering and Real Property Management, NASA.

August 29 Second Committee Meeting; Interview with Dwight A. Beranek, U.S. Army Corps of Engineers; Joe Gott, P.E., Naval Facilities Engineering Command; William C. Stamper, Department of Health and Human Services; Stan Kaczmarczyk, General Services Administration; Mike Carosotto, U.S. Coast Guard; Phil Dalby and Pete O'Konski, U.S. Department of Energy.

Briefing by Judith W. Passwaters, Global Director, Facilities Services and Real Estate, E.I. duPont de Nemours & Company; John Palguta, Vice President for Policy and Research, The Partnership for Public Service.

2006

April 11 Web briefing by David M. Hammond, RLA, Office of Civil Engineering, U.S. Coast Guard.

Appendix C

Executive Summary from *Stewardship of Federal Facilities*

Since its establishment in 1789, the federal government has constructed and acquired buildings, other facilities, and their associated infrastructures to support the conduct of public policy, defend the national interest, and provide services to the American public. Today, the federal facilities inventory comprises more than 500,000 buildings and structures, as well as the power plants, utility distribution systems, roads, and other infrastructure required to support them. Federal facilities are located in all 50 states, U.S. territories, and more than 160 foreign countries. They span decades, sometimes centuries, of building design and construction technologies, support a myriad of government functions, and represent the investment of more than 300 billion tax dollars.[1]

Federal facilities embody significant investments and resources and therefore constitute a portfolio of public assets. These buildings and structures project an image of American government at home and abroad, contribute to the architectural and socioeconomic fabric of their communities, and support the organizational and individual performance of federal employees conducting the business of government. These assets must be well maintained to operate adequately and cost effectively, to protect their functionality and quality, and to provide a safe, healthy, productive environment for the American public, elected officials, federal employees, and foreign visitors who use them every day.

Despite the historic, architectural, cultural, and functional importance of, and the economic investment in, federal facilities, studies by the General Accounting Office (GAO) and other federal government agencies indicate that the physical

---

[1]As of fiscal year 1996, federal agencies reported $215.5 billion of investment in structures and facilities and almost $82 billion of construction in progress (GAO, 1998).

condition of this portfolio of public assets is deteriorating. Many necessary repairs were not made when they would have been most cost effective and have become part of a backlog of deferred maintenance. In addition, a large proportion of federal facilities are more than 40 years old. As wear and tear on buildings and their systems increases, the need for maintenance and repair to sustain their performance and functionality also increases. Federal agency program managers, the GAO, and research organizations have all reported that the funding allocated for the repair and maintenance of federal facilities is insufficient.

Although there is no single, agreed upon guideline to determine how much money is, in fact, necessary to maintain public buildings, a 1990 report of the National Research Council, *Committing to the Cost of Ownership: The Maintenance and Repair of Public Buildings*, did recommend that "an appropriate budget allocation for routine M&R [maintenance and repair] for a substantial inventory of facilities will typically be in the range of 2 to 4 percent of the aggregate current replacement value of those facilities" (NRC, 1990). This guideline has been widely quoted in the facilities management literature. During the course of the present study, federal agency representatives indicated that the funding they receive for maintenance and repair of their agencies' facilities is less than 2 percent of the aggregate current replacement value of their facilities inventories.

In an environment of inadequate and declining resources, federal facilities program managers face a number of challenges:

- Extending the useful life of aging facilities.
- Altering or retrofitting facilities to consolidate space or accommodate new functions and technologies.
- Meeting evolving facility-related standards for safety, environmental quality, and accessibility.
- Maintaining or disposing of excess facilities created through military base closures and realignments, downsizing, or changing demographics.
- Finding innovative ways and technologies to maximize limited resources.

To help federal agencies meet these challenges and optimize available resources, the sponsoring agencies of the Federal Facilities Council requested that the National Research Council review current federal practices for planning, budgeting, and implementing facility maintenance and repair programs and (1) develop a methodology and rationale federal facilities program managers can use for the systematic formulation and justification of facility maintenance and repair budgets; (2) investigate the role of technology in performing automated condition assessments; and (3) identify staff capabilities necessary to perform condition assessments and develop maintenance and repair budgets.

The study was conducted under the auspices of the Board on Infrastructure and the Constructed Environment by a committee of recognized experts in budgeting, facilities operations and maintenance, decision science, engineering economics, and building and facilities technology. Many of the committee members have professional experience with the management of large facilities portfolios. In addition, they were assisted by representatives of federal agencies, private sector organizations, and individuals who provided information on current budgeting, financial, maintenance, and building engineering practices in the federal government and the private sector.

Throughout this study, the committee was hampered by a lack of published data related to federal facilities inventories, programs, and practices. Accurate counts of basic items, such as the total number of federal facilities, the age of facilities, and expenditures for maintenance and repair, were not available. (This issue is addressed in the study's findings and recommendations.) The committee also found that current maintenance and repair budgeting procedures, definitions, and accounting have advanced little since 1990. For information on the physical condition of federal facilities, maintenance and repair budgeting, condition assessment practices, deferred maintenance, and related topics, the committee relied heavily on reports of the GAO, briefings by federal agency program managers, and personal experience.

The committee began task 1 with the idea that it could develop a methodology for the systematic formulation of maintenance and repair budgets. However, the current state of practice, the general lack of data, and the lack of research results, in particular, precluded the development of a methodology per se. Instead, the committee identified methods, principles, and strategies that, if implemented, can be used to develop a methodology for the systematic formulation of maintenance and repair budgets in the future. In approaching task 2, the committee reviewed federal agency condition assessment practices and the role of technology in developing automated condition assessments. The committee found that existing sensor and microprocessor technologies have the potential to monitor and manage a range of building conditions and environmental parameters, but, for economic and other reasons, they have not been widely deployed. The committee also reviewed staff capabilities necessary to the performance of condition assessments and the development of maintenance and repair budgets (task 3). The committee found that both require adequate training for staff to foster effective decision making in facilities management, condition assessments, and maintenance and repair budgeting.

Federal government standards for internal oversight and control require that agencies safeguard all of the assets entrusted to them. This report seeks to foster accountability for the stewardship (i.e., responsible care) of federal facilities at all levels of government. The committee's findings and recommendations are presented below.

FINDINGS

Finding 1. Based on the information available to the committee, the physical condition of the federal facilities portfolio continues to deteriorate, and many federal buildings require major repairs to bring them up to acceptable quality, health, and safety standards.

Finding 2. The underfunding of facilities maintenance and repair programs is a persistent, long-standing problem. Federal operating and oversight agencies report that agencies have excess, aging facilities and insufficient funds to maintain, repair, or update them. Information provided to the committee indicated that agencies are receiving less than 2 percent of the aggregate current replacement value of their facilities inventories for maintenance and repair.

Finding 3. Federal government processes and practices are generally not structured to provide for effective accountability for the stewardship (i.e., responsible care) of federal facilities.

Finding 4. Buildings and facilities are durable assets that contribute to the effective provision of government services and the fulfillment of agency missions. However, the relationship of facilities to agency missions has not been recognized adequately in federal strategic planning and budgeting processes.

Finding 5. Maintenance and repair expenditures generally have less visible or less measurable benefits than other operating programs. Facilities program managers have found it difficult to make compelling arguments to justify these expenditures to public officials, senior agency managers, and budgeting staff.

Finding 6. Budgetary pressures on federal agency managers encourage them to divert potential maintenance and repair funds to support current operations, to meet new legislative requirements, or to pay for operating new facilities coming on line.

Finding 7. It is difficult, if not impossible, to determine how much money the federal government as a whole appropriates and spends for the maintenance and repair of federal facilities because definitions and calculations of facilities-related budget items vary, as do methodologies for developing budgets and accounting and reporting systems for tracking maintenance and repair expenditures.

Finding 8. There is evidence that some agencies own and are responsible for more facilities than they need to support their missions or than they can maintain with current or projected budgets.

Finding 9. Federal facilities program managers are being encouraged to be more business-like and innovative, but current management, budgeting, and financial processes have disincentives and institutional barriers to cost-effective facilities management and maintenance practices.

Finding 10. Performance measures to determine the effectiveness of maintenance and repair expenditures have not been developed within the federal government. Thus, it is difficult to identify best practices for facilities maintenance and repair programs across or within federal agencies.

Finding 11. Based on the information available to the committee, federal condition assessment programs are labor intensive, time consuming, and expensive. Agencies have had limited success in making effective use of the data they gather for timely budget development or for the ongoing management of facilities.

Finding 12. Organizational downsizing has forced facilities program managers to look increasingly to technology solutions to provide facilities-related data for decision making and for performing condition assessments.

Finding 13. Existing sensor and microprocessor technologies have the potential to monitor and manage a range of building conditions and environmental parameters, but, for economic and other reasons, they have not been widely deployed.

Finding 14. Training for staff is a key component of effective decision making, condition assessments, and the development of maintenance and repair budgets.

Finding 15. Only a limited amount of research has been done on the deterioration/failure rates of building components or the nonquantitative implications of building maintenance (or lack thereof). This research is necessary to identify effective facilities management strategies for achieving cost savings, identifying cost avoidances, and providing safe, healthy, productive work environments.

Finding 16. Greater accountability for the stewardship of facilities is necessary at all levels of the federal government. Accountability includes being held responsible for the condition of facilities and for the allocation, tracking, and effective use of maintenance and repair funds.

The committee recommends that the government take the following actions (which are not in any particular order of priority).

RECOMMENDATIONS

Recommendation 1. The federal government should plan strategically for the maintenance and repair of its facilities in order to optimize available resources, maintain the functionality and quality of federal facilities, and protect the public's investment. A recommended strategic framework of methods, practices, and strategies for the proactive management and maintenance of the nation's public assets is summarized on Figure ES-1 (Findings 1 and 2).

Recommendation 2. The government should foster accountability for the stewardship of federal facilities at all levels. Facilities program managers at the agency level should identify and justify the resources necessary to maintain facilities effectively and should be held accountable for the use of these resources (Findings 1, 2, 3, and 16).

Recommendation 3. At the executive level, an advisory group of senior level federal managers, other public sector managers, and representatives of the nonprofit and private sectors should be established to develop policies and strategies to foster accountability for the stewardship of facilities and to allocate resources strategically for their maintenance and repair (Findings 1, 2, 3, and 16).

Recommendation 4. Facility investment and management should be directly linked to agency mission. Every agency's strategic plan should include a facilities component that links facilities to agency mission and establishes a basis and rationale for maintenance and repair budget requests (Finding 4).

Recommendation 5. The government should adopt more standardized budgeting and cost accounting techniques and processes to facilitate tracking of maintenance and repair funding requests, allocations, and expenditures and reflect the total costs of facilities ownership. The committee developed an illustrative template as shown in Figure ES-2 (Findings 3, 5, 6, 7, and 16).

APPENDIX C

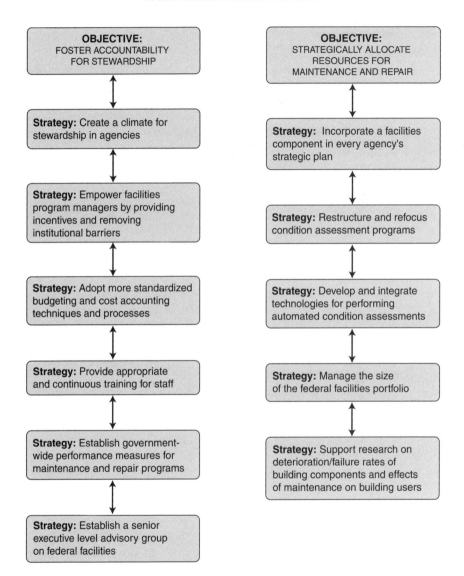

FIGURE ES-1 Strategic framework for the maintenance and repair of federal facilities.

Facilities Management-Related Activities	Included in 2%-4% Benchmark	Funding Category and Comments
A Routine Maintenance, Repairs, and Replacements • recurring, annual maintenance and repairs including maintenance of structures and utility systems (including repairs under a given $ limit, e.g., $150,000 to $500,000 exclusive of furniture and office equipment) • roofing, chiller/boiler replacement, electrical/lighting, etc. • preventive maintenance • preservation/cyclical maintenance • deferred maintenance backlog • service calls	Yes	Annual operating budget
B Facilities-Related Operations • custodial work (i.e., services and cleaning) • utilities (electric, gas, etc./plant operations) • snow removal • waste collection and removal • pest control • security services • grounds care • parking • fire protection services	No	Annual operating budget
C Alterations and Capital Improvements • major alterations to subsystems, (e.g., enclosure, interior, mechanical, electrical expansion) that change the capacity or extend the service life of a facility • minor alterations (individual project limit to be determined by agency $50,000 to $1 million)	No	Various funding sources, including no year, project-based allocations such as revolving funds, carryover of unobligated funds, funding resulting from cost savings or cost avoidance strategies
D Legislatively Mandated Activities • improvements for accessibility, hazardous materials removal, etc.	No	Various sources of funding
E New Construction and Total Renovation Activities	No	Project-based allocations separate from operations and maintenance budget. Should include a life-cycle cost analysis prior to funding
F Demolition Activities	No	Various sources of funding

FIGURE ES-2 Illustrative template to reflect the total costs of facilities ownership.

Recommendation 6. Government-wide performance measures should be established to evaluate the effectiveness of facilities maintenance and repair programs and expenditures (Finding 10).

Recommendation 7. Facilities program managers should be empowered to operate in a more businesslike manner by removing institutional barriers and providing incentives for improving cost-effective use of maintenance and repair funds. The carryover of unobligated funds and the establishment of revolving funds for nonrecurring maintenance needs should be allowed if they are justified (Findings 3 and 9).

Recommendation 8. Long-term requirements for maintenance and repair expenditures should be managed by reducing the size of the federal facilities portfolio. New construction should be limited, existing buildings should be adapted to new uses, and the ownership of unneeded buildings should be transferred to other public or private organizations. Facilities that are functionally obsolete, are not needed to support an agency's mission, are not historically significant, and are not suitable for transfer or adaptive reuse should be demolished whenever it is cost effective to do so (Findings 2, 8, and 16).

Recommendation 9. Condition assessment programs should be restructured to focus first on facilities that are critical to an agency's mission; on life, health, and safety issues; and on building systems that are critical to a facility's performance. This will optimize available resources, provide timely and accurate data for formulating maintenance and repair budgets, and provide critical information for the ongoing management of facilities (Findings 4 and 11).

Recommendation 10. The government should provide appropriate and continuous training for staff that perform condition assessments and develop and review maintenance and repair budgets to foster informed decision making on issues related to the stewardship of federal facilities and the total costs of facilities ownership (Findings 14 and 16).

Recommendation 11. The government and private industry should work together to develop and integrate technologies for performing automated facility condition assessments and to eliminate barriers to their deployment (Findings 11, 12, and 13).

Recommendation 12. The government should support research on the deterioration/failure rates of building components and the nonquantitative effects of building maintenance (or lack thereof) in order to develop

quantitative data that can be used for planning and implementing cost-effective maintenance and repair programs and strategies and for better understanding the programmatic effects of maintenance on mission delivery and on building users' health, safety, and productivity (Findings 12 and 15).

REFERENCES

GAO (General Accounting Office). 1998. Deferred Maintenance Reporting: Challenges to Implementation. Report to the Chairman, Committee on Appropriations, U.S. Senate. AIMD-98-42. Washington, D.C.: Government Printing Office.

NRC (National Research Council). 1990. Committing to the Cost of Ownership: Maintenance and Repair of Public Buildings. Building Research Board, National Research Council. Washington, D.C.: National Academy Press.

Appendix D

Executive Summary from *Outsourcing Management Functions*

In this study *outsourcing* is defined as the organizational practice of contracting for services from an external entity while retaining control over assets and oversight of the services being outsourced. In the 1980s, a number of factors led to a renewed interest in outsourcing. For private sector organizations, outsourcing was identified as a strategic component of business process reengineering—an effort to streamline an organization and increase its profitability. In the public sector, growing concern about the federal budget deficit, the continuing long-term fiscal crisis of some large cities, and other factors accelerated the use of privatization[1] measures (including outsourcing for services) as a means of increasing the efficiency of government.

The literature on business management has been focused on the reengineering of business processes in the context of the financial, management, time, and staffing constraints of private enterprise. The underlying premises of business process reengineering are (1) the essential areas of expertise, or core competencies, of an organization should be limited to a few activities that are central to its current focus and future profitability, or bottom line, and (2) because managerial time and resources are limited, they should be concentrated on the organization's core competencies. Additional functions can be retained within the organization, or in-house, to keep competitors from learning, taking over, bypassing, or eroding the organization's core business expertise. Routine or noncore elements of the business can be contracted out, or outsourced, to external entities that specialize in those services.

---

[1] Privatization has been defined as any process aimed at shifting functions and responsibilities, in whole or in part, from the government to the private sector.

Public-sector organizations, in contrast, have no bottom line comparable to the profitability of a business enterprise. The missions of governmental entities are focused on providing services related to public health, safety, and welfare; one objective is to do so cost effectively, rather than profitably. Thus, public practices are often very different from private-sector practices. They entail different risks, different operating environments, and different management systems.

Private corporations and the federal government have invested billions of dollars in facilities and infrastructure to support the services and activities necessary to fulfill their respective businesses and missions. Until the corporate downsizings of the 1980s, owners of large inventories of buildings usually maintained in-house facilities engineering organizations responsible for design, construction, operations, and project management. These engineering organizations were staffed by hundreds, sometimes thousands, of architects and engineers. In the United States during the last 20 years, almost all of these engineering organizations have been reorganized, sometimes repeatedly, as a result of business process reengineering. Some organizations are still restructuring their central engineering organizations, shifting project responsibilities to business units or operating units, and outsourcing more work to external organizations.

Studies have found that many companies are uncertain about the appropriate size and role of their in-house facilities engineering organizations. Reorganizations sometimes leave owners inadequately structured to develop and execute facility projects. In many organizations, the technical competence necessary to develop the most appropriate project to meet a business need has been lost, along with the competency to execute the project effectively. Even though many owner organizations recognize that the skills required on the owner's side to manage projects have changed dramatically, they are doing little to address this issue.

Federal agencies are experiencing changes similar to those affecting private sector owner organizations. A survey by the Federal Facilities Council found that by 1999, in nine federal agencies, in-house facilities engineering staffs had been reduced by an average of 50 percent. The loss of expertise reflected in this statistic is compounded by the fact that procurement specialists, trained primarily in contract negotiation and review rather than in design and construction, are playing increasingly greater roles in facility acquisitions.

Outsourcing is not new to federal agencies. The government has contracted for facility planning, design, and construction services for decades. Recently, however, in response to executive and legislative initiatives to reduce the federal workforce, cut costs, improve customer service, and become more businesslike, federal agencies have begun outsourcing some management functions for facility acquisitions. The reliance on nonfederal entities to provide management functions for federal facility acquisitions has raised concerns about the level of control, management responsibility, and accountability being transferred to nonfederal service providers. Outsourcing management functions has also raised concerns about some agencies' long-term ability to plan, guide, oversee, and evaluate facility acquisitions effectively.

To address these concerns, the sponsoring agencies of the Federal Facilities Council requested that the National Research Council (NRC) develop a guide, or "road map," to help federal agencies determine which management functions for planning, design, and construction-related services may be outsourced. In carrying out this charge, the NRC committee appointed to prepare this report was asked to (1) assess recent federal experience with the outsourcing of management functions for planning, design, and construction services; (2) develop a technical framework and methodology for implementing a successful outsourcing program; (3) identify measures to determine performance outcomes; and (4) identify the organizational core competencies necessary for effective oversight of outsourced management functions while protecting the federal interest.

DETERMINING WHICH MANAGEMENT FUNCTIONS MAY BE OUTSOURCED BY FEDERAL AGENCIES

The committee reviewed federal legislation and policies related to inherently governmental functions—a critical determinant of which activities federal agencies can and cannot outsource. An inherently governmental function is defined as one that is so intimately related to the public interest that it must be performed by government employees. An activity not inherently governmental is defined as commercial. The committee concluded that, although design and construction activities are commercial and may be outsourced, management functions cannot be clearly categorized.

In the facility acquisition process, an owner's role is to establish objectives and to make decisions on important issues. Management functions, in contrast, include the ministerial tasks necessary to accomplish the task. Based on a review of federal regulations, the committee concluded that inherently governmental functions related to facility acquisitions include making a decision (or casting a vote) pertaining to policy, prime contracts, or the commitment of government funds. None of these can be construed as ministerial functions. The distinction between activities that are inherently governmental and those that are commercial, therefore, is essentially the same as the distinction between ownership and management functions.

Using Section 7.5 of the Federal Acquisition Regulations as a basis, the committee developed a two-step threshold test to help federal agencies determine which management functions related to facility acquisitions should be performed by in-house staff and which may be considered for outsourcing to external organizations. The first step is to determine whether the function involves decision making on important issues (ownership) or ministerial or information-related services (management). In the committee's opinion, ownership functions should be performed by in-house staff and should not be outsourced.

For activities deemed to be management functions, the second step of the analysis is to consider whether outsourcing the management function might unduly compromise one or more of the agency's ownership functions. If outsourcing of a management function would unduly compromise the agency's ownership role, then it should be considered a "quasi"-inherently governmental function and should not be outsourced.

Figure ES-1 is a decision framework developed by the committee for federal agencies considering outsourcing management functions for facility acquisitions. This framework recognizes the constraints of inherently governmental functions and incorporates the committee's two-step threshold for identifying ownership functions that should be performed by in-house staff and management functions that can be considered for outsourcing. The decision framework is not intended to generate definitive recommendations about which management functions may or may not be outsourced or in what combination. The decision framework is a tool to assist decision makers in analyzing their organizational strengths and weaknesses, assessing risk in specific areas based on a project's stature and sensitivity, and, at a fundamental level, questioning whether or not a management function can best be performed by in-house staff or by an external organization.

The line between inherently governmental functions and commercial activities or between ownership and management functions can be very fine. Distinguishing between them can be difficult and may require a case-by-case analysis of many facts and circumstances.

FEDERAL EXPERIENCE WITH THE OUTSOURCING OF MANAGEMENT FUNCTIONS

The authoring committee received briefings from several federal agencies and developed and distributed a questionnaire to sponsoring agencies of the Federal Facilities Council to solicit information on their experiences with outsourcing in general and outsourcing of management functions in particular. Seven of the 13 agencies that responded to the questionnaire had outsourced some management functions for planning, design, and construction-related activities. The primary factors cited for outsourcing management functions were lack of in-house expertise and staff shortages (54 percent of responses combined); savings on project delivery time (15 percent); and other factors, including statutory requirements (15 percent). None of the seven agencies cited cost effectiveness or deliberate downsizing as a factor in the decision to outsource management functions. Three of the seven had outsourced management functions to other federal agencies. Their experiences varied and no trends could be determined. Agencies' experiences with outsourcing management functions to the private sector were also varied, and, again, no trends could be discerned.

APPENDIX D

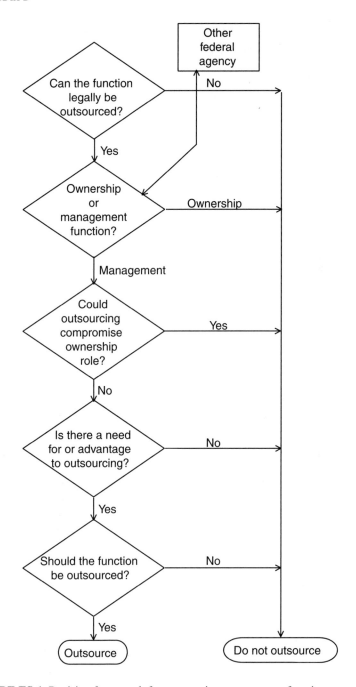

FIGURE ES-1 Decision framework for outsourcing management functions.

ORGANIZATIONAL CORE COMPETENCIES

At any one time, a federal agency may be responsible for managing several dozen to several hundred individual projects in various stages of planning, design, and construction. In some cases, agencies acquire facilities with the intent of owning and managing them directly. In other cases, agencies only require the use of facilities and may use a procuring entity to represent the government-as-owner in the acquisition process. A few agencies provide facilities for other agencies and organizations as a key component of their missions.

Core competencies constitute an organization's essential area of expertise and skill base. Unless a federal agency's mission is to provide facilities, facility acquisition and management are not core functions (i.e., facilities are not the mission but support accomplishment of the mission). However, when acquiring facilities, federal agencies assume an ownership responsibility as a steward of the public's investment. The requirements that a federal agency be accountable for upholding public policy and for committing public resources are indivisible. This combination of responsibilities requires that any federal agency that acquires facilities have the in-house capabilities to translate its mission needs directly into program definitions and project specifics and otherwise act in a publicly responsive and accountable manner. Other organizational core competencies required to direct and manage specific projects vary, depending on the agency's role as owner, user, or provider of a facility.

IMPLEMENTING A SUCCESSFUL OUTSOURCING PROGRAM

Once a decision has been made to outsource some or all management functions for facility acquisitions, the agency should clearly define the roles and responsibilities of all of the entities involved. The committee recommends that federal agencies establish and apply a responsibilities-and-deliverables matrix similar to the example shown in Figure ES-2 to help eliminate overlapping responsibilities, ensure accountability, and ensure that, as problems arise, solutions are managed effectively.

DETERMINING PERFORMANCE OUTCOMES

A key element of an organization's decision making is measuring the effectiveness of those decisions, both qualitatively and quantitatively. When management functions for facility acquisitions are outsourced, the principal measures of effectiveness of the entire effort and of individual projects should relate to cost, schedule, and safety of the projects, as well as the functionality and overall quality of the acquired facilities. If baseline levels of service already exist or can be developed empirically, comparing the metrics and determining how well the outsourcing effort meets the basic level of expectation should be straightforward.

APPENDIX D

RESPONSIBILITIES-AND-DELIVERABLES MATRIX

	User Management	Owner Management	Owner Project Manager	Outsourced Project Management	(A – E)	Construction Contractor	Specialty Contractors
Programming Phase							
Project request	A	P	S	S			
Deliverables/responsibilities package			A	P			
Conceptual Planning Phase							
Architect-engineer contracts			A		P		
Detailed requirements	R	A	P	S	S		
Design Phase							
Conceptual and schematic designs	R	C	A	C	P		
Permits			A	C	P		
Design development	A	A	A	C	P		
Construction documents			C	A	P		S
Procurement Phase							
List of bidders and requests for proposals			A	C	P		
Proposals (submitted)			A	C	S	P	P
Contract for construction			A	P	S	S	S
Construction Phase							
Construction permits			A	C		P	
Construction management			C	A	S	P	P
Construction work			C	A	S	P	P
Final payment (construction complete)		A	A	C	S	P	P
Start-up Phase							
Equipment installation				C	A	S	P
Move administration			P	S	S		S
Final acceptance	A	A	C	P		S	
Closeout Phase			P	S	S	S	S

FIGURE ES-2 Example of a responsibilities-and-deliverables matrix. P, primary responsibility; A, approve (signing of approval); C, concurrence; R, reviews (no response required); and S, support (uses own resources).

If no baseline exists, one should be developed to ensure effective performance measurement.

Individual performance measures should be developed by the agencies that will use them and should not be prescribed by higher levels of government. Although it is entirely appropriate that operational guidance requiring the use of performance measures to be addressed be promulgated government-wide (e.g., Government Performance and Results Act) and to specify what these measures should address, the parties actually responsible for the provision of a service are in the best position to determine what constitutes good performance. Any agency that decides to outsource management functions for planning, design, and construction services should be prepared to develop and apply meaningful, measurable performance measures to determine if it is meeting its stewardship responsibilities.

FINDINGS AND RECOMMENDATIONS

The primary objective of this study is to develop a guide that federal agencies can use in the initial stages of decision making concerning the outsourcing of management functions for planning, design, and construction-related services. Agencies will have to expand and extend the guidance in this report and tailor it to their individual circumstances. By using the decision framework, by noting the findings, and by following the recommendations presented below, the committee believes federal agencies will be in a stronger position to formulate rational, business-like judgments in the public interest concerning the outsourcing of management functions for planning, design, and construction-related services.

Findings

Finding. Each federal agency involved in acquiring facilities is accountable to the U.S. government and its citizens. Each agency is responsible for managing its facilities projects and programs effectively. Responsibility for stewardship cannot be outsourced.

Finding. The outsourcing of management functions for planning, design, and construction-related services by federal agencies is a strategic decision that should be considered in the context of an agency's long-term mission.

Finding. The outsourcing of management functions for planning, design, and construction services has been practiced by some federal agencies for years. Management functions have been outsourced either to other federal agencies or the private sector. The outcomes of these efforts have varied widely, from failure to success.

Finding. At different times, an agency may fill one or more of the role(s) of owner, user, or provider of facilities.

Finding. Key factors in determining successful outcomes of outsourcing decisions include clear definitions of the scope and objectives of the services required at the beginning of the acquisition process and equally clear definitions of the roles and responsibilities of the agency. Owners and users need to provide leadership; define scope, goals, and objectives; establish performance criteria for evaluating success; allocate resources; and provide commitment and stability for achieving the goals and objectives.

Finding. Program scope, definition, and budget decisions are inherently the responsibilities of owners/users and should not be outsourced. However, assistance in discharging these responsibilities may have to be obtained by contracting for services from other federal agencies or the private sector.

Finding. The successful outsourcing of management functions by federal agencies requires competent in-house staff with a broad range of technical, financial, procurement, and management skills and a clear understanding of the agency's mission and strategic objectives.

Finding. Performance measures are necessary to assess the success of any outsourcing effort.

Finding. Because federal facilities vary widely, and because a wide range of new and evolving project delivery systems have inherently different levels of risk and management requirements, no single approach or set of organizational core competencies for the acquisition of federal facilities applies to all agencies or situations.

Finding. The organizational core competencies necessary to oversee the outsourcing of management functions for projects and/or programs need to be actively nurtured over the long term by providing opportunities for staff to obtain direct experience and training in the area of competence. The necessary skills will, in part, be determined by the role(s) the agency fills as owner, user, and/or provider of facilities.

Recommendations

Recommendation. A federal agency should analyze the relationship of outsourcing decisions to the accomplishment of its mission before outsourcing management functions for planning, design, or construction services. Outsourcing for services and functions should be integrated into an overall strategy for achieving the agency's mission, managing resources, and obtaining best value or best performance for the resources expended. Outsourcing of management functions should not be used solely as a short-term expedient to limit spending or reduce the number of in-house personnel.

Recommendation. Federal agencies should first determine their role(s) as owners, users, and/or providers of facilities and then determine the core competencies required to effectively fulfill these role(s) in overseeing the outsourcing of management functions for planning, design, and construction services.

Recommendation. Once a decision has been made to outsource some or all management functions, a responsibilities-and-deliverables matrix should be established to help eliminate overlapping responsibilities, provide accountability, and ensure that, as problems arise, solutions are managed effectively.

Recommendation. Agencies that outsource management functions for planning, design, and construction services should regularly evaluate the effectiveness of the outsourcing effort in relation to accomplishment of the agency's mission.

Recommendation. Agencies should establish performance measures to assess accomplishments relative to the objectives established for the outsourcing effort and, at a minimum, address cost, schedule, and quality parameters.

Recommendation. Owner/user agencies should retain a sufficient level of technical and managerial competency in-house to act as informed owners and/or users when management functions for planning, design, and construction services are outsourced.

Recommendation. Provider agencies should retain a sufficient level of planning, design, and construction management activity in-house to ensure that they can act as competent providers of planning, design, and construction management services.

Recommendation. Agencies should provide training for leaders and staff responsible for technical, procurement, financial, business, and managerial functions so that they can effectively oversee the outsourcing of management functions for planning, design, and construction.

Recommendation. Interagency coordination, cooperation, collaboration, networking, and training should be increased to encourage the use of best practices and improve life-cycle cost effectiveness in the delivery of federal facilities.

Appendix E

Executive Summary from *Investments in Federal Facilities*

Federal facilities investments are matters of public policy. The facilities acquired by the federal government provide a means to produce and distribute public goods and services to 280 million Americans, create jobs, strengthen the national economy, and support the missions of federal departments and agencies, including the defense and security missions. Such investments also support policies for public transportation, urban revitalization, and historic preservation, among others.

Hundreds of billions of dollars have been invested in federal facilities and their associated infrastructure. As of September 2000, the federal government owned or leased 3.3 billion square feet of space worldwide, distributed across more than 500,000 facilities conservatively valued at $328 billion. Annually, it spends upwards of $21 billion for the acquisition and renovation of facilities; approximately $4.5 billion to power, heat, and cool its buildings; and more than $500 million for water and waste disposal. Additional expenditures for facilities maintenance, repair, renewal, demolition, and security upgrades probably amount to billions of dollars per year but are not readily identifiable under the current budget structure.

Despite the magnitude of this ongoing investment, federal facilities continue to deteriorate, backlogs of deferred maintenance continue to increase, and excess, underutilized, and obsolete facilities continue to consume limited resources. Many departments and agencies have the wrong facilities, too many or not enough facilities, or facilities that are poorly sited to support their missions. Such facilities constitute a drain on the federal budget in actual costs and in foregone opportunities to invest in other public resources and programs.

On January 30, 2003, the U.S. General Accounting Office (GAO) designated federal real property as a government-wide high-risk area[1] because current trends "have multibillion dollar cost implications and can seriously jeopardize mission accomplishment" and because "federal agencies face many challenges securing real property due to the threat of terrorism." It declared that "current structures and processes may not be adequate to address the problems," so that "a comprehensive, integrated transformation strategy" may be required.

PRINCIPLES AND POLICIES FOR FACILITIES INVESTMENTS AND MANAGEMENT

As the committee reviewed the types of analyses, the processes, and the decision-making environments that private sector and other organizations use for facilities investments and management, it focused on identifying principles and policies used by best-practice organizations, as defined by the committee. The committee found that, in matters of facilities investment and management, best-practice organizations do the following:

Principle/Policy 1. Establish a framework of procedures, required information, and valuation criteria that aligns the goals, objectives, and values of their individual decision making and operating groups to achieve the organization's overall mission; create an effective decision-making environment; and provide a basis for measuring and improving the outcomes of facilities investments. The components of the framework are understood and used by all leadership and management levels.

Principle/Policy 2. Implement a systematic facilities asset management approach that allows for a broad-based understanding of the condition and functionality of their facilities portfolios—as distinct from their individual projects—in relation to their organizational missions. Best-practice organizations ensure that their facilities and infrastructure managers possess both the technical expertise and the financial analysis skills to implement a portfolio-based approach.

Principle/Policy 3. Integrate facilities investment decisions into their organizational strategic planning processes. Best-practice organizations evaluate facilities investment proposals as mission enablers rather than solely as costs.

---

[1] GAO's high-risk update is provided at the start of each new Congress. The reports are intended to help the new Congress "focus its attention on the most important issues and challenges facing the federal government." (GAO, 2003f)

Principle/Policy 4. Use business case analyses to rigorously evaluate major facilities investment proposals and to make transparent a proposal's underlying assumptions; the alternatives considered; a full range of costs and benefits; and the potential consequences for their organizations.

Principle/Policy 5. Analyze the life-cycle costs of proposed facilities, the life-cycle costs of staffing and equipment inherent to the proposal, and the life-cycle costs of the required funding.

Principle/Policy 6. Evaluate ways to disengage from, or exit, facilities investments as part of the business case analysis and include disposal costs in the facilities life-cycle cost to help select the best solution to meet the requirement.

Principle/Policy 7. Base decisions to own or lease facilities on the level of control required and the planning horizon for the function, which may or may not be the same as the life of the facility.

Principle/Policy 8. Use performance measures in conjunction with both periodic and continuous long-term feedback to evaluate the results of facilities investments and to improve the decision-making process itself.

Principle/Policy 9. Link accountability, responsibility, and authority when making and implementing facilities investment decisions.

Principle/Policy 10. Motivate employees as individuals and as groups to meet or exceed accepted levels of performance by establishing incentives that encourage effective decision making and reward extraordinary performance.

ADAPTING THE PRINCIPLES AND POLICIES TO THE FEDERAL OPERATING ENVIRONMENT

Adapting the aforementioned principles and policies for facilities investments for use by the federal government requires consideration of and compensation for a number of special aspects of the federal operating environment. These aspects include the goals and missions of the federal government, its departments, and agencies; the organizational structure and decision-making environment; the nature of federal facilities investments; and the annual budget process and its attendant procedures. They are described more fully in Chapters 1 and 6 [of Investments in Federal Facilities].

Despite the inherent differences, the committee's overall conclusion is that aspects of all of the identified principles and policies used by best-practice organizations can be adapted in varying form to the federal operating environment. It has therefore made recommendations to aid in developing an overall framework based on suitable adaptations of the identified principles and policies.

The committee also concluded that there is no single solution from the private sector that can be applied to all issues related to federal facilities investment and management, nor should there be an expectation that one will be found. The committee points to the number of missions and the variation in size, resources, culture, and political support of the many federal agencies with facilities-related responsibilities and urges all involved not to attempt to create one-size-fits-all solutions to different problems.

Instead, the committee recommends that efforts be made to concurrently and collaboratively develop top-down and bottom-up approaches while keeping in mind differences among various agency missions and cultures as well as similarities in many specifics of facility investment and management. Varying practices within common principles and policies should be expected.

RECOMMENDATION 1. The federal government should adopt a framework of procedures, required information, and valuation criteria for federal facilities investment decision making and management that incorporates all of the principles and policies enumerated by this committee.

Implementation of a framework that incorporates the identified principles and policies will align the goals, objectives, and values of individual federal decision-making and operating groups with overall missions; create an effective decision-making environment; and provide a basis for measuring and improving the outcomes of federal facilities investments. Because such a framework represents a significant departure from current operating procedures, it may be advisable to establish one or more pilot projects. A small government agency with a diverse portfolio of facilities might provide the environment in which to test the application of the committee's recommendations.

RECOMMENDATION 2(a). Each federal department and agency should update its facilities asset management program to enable it to make investment and management decisions about individual projects relative to its entire portfolio of facilities.

Federal departments and agencies have begun implementing facilities asset management approaches that allow for a broad-based understanding of the condition and functionality of their facilities portfolios. An updated approach should incorporate life-cycle decision making that accounts for all the inherent operating costs (i.e., facilities, staffing, equipment, and information technologies); accurate databases; condition assessments; performance measures; feedback processes; and appropriately adapted business practices.

RECOMMENDATION 2(b). Each federal department and agency should ensure it has the requisite technical and business skills to implement a facilities asset management approach by providing specialized training for its incumbent facilities asset management staff and by recruiting individuals with these skills.

Most federal departments and agencies currently have staff with the requisite technical skills to implement asset management approaches. Less likely to be found are facilities management staff also versed in financial theory, practices, and management. Departments and agencies should provide their incumbent facilities asset management staff with training in business concepts such as financial theory and analysis. Training can be in the form of coursework, seminars, rotational assignments, and other appropriate methods. As job vacancies occur in facilities management operating groups, departments and agencies should seek to recruit and hire staff with the requisite technical and business skills.

RECOMMENDATION 2(c). To facilitate the alignment of each department's and agency's existing facilities portfolios with its missions, Congress and the administration should jointly lead an effort to consolidate and streamline government-wide policies, regulations, and processes related to facilities disposal, which would encourage routine disposal of excess facilities in a timely manner.

Eighty-one separate policies applicable to the disposal of federal facilities have been identified. These include agency-specific legislative mandates and directives and government-wide socioeconomic and environmental policies. The number of policies related to facilities disposal hinders government-wide efforts to expeditiously dispose of unneeded facilities in response to changing requirements.

RECOMMENDATION 2(d). For departments and agencies with many more facilities than are needed for their missions—the Departments of Defense, Energy, State, and Veterans Affairs, the General Services Administration, and possibly others—Congress and the administration should jointly consider implementing extraordinary measures like the process used for military base realignment and closure (BRAC), modified as required to reflect actual experience with BRAC.

Federal agencies are incurring significant costs by operating and maintaining facilities they no longer need to support today's missions. The Department of Defense (DoD) alone estimates it spends $3 to $4 billion each year maintaining excess facilities. The lack of alignment between a department's or agency's mission and its facilities portfolio, coupled with the cost of operating and maintaining excess facilities, may require extraordinary measures to effect improvement, such as the BRAC process used for closing DoD facilities. The government as a whole and the DoD in particular have 15 years of experience and lessons from BRAC. Such lessons can be used to make adjustments to the process to improve it and adapt it to other departments and agencies, as appropriate.

RECOMMENDATION 3. Each federal department and agency should use its organizational mission as guidance for facilities investment decisions and should then integrate facilities investments into its organizational strategic planning processes. Facilities investments should be evaluated as mission enablers, not solely as costs.

Organizational strategic planning that does not include facilities considerations up front fails to account for a potentially substantial portion of the total cost of a program or initiative. Integrating facilities considerations into evaluations of strategic planning alternatives will provide decision makers with better information about the total long-term costs, considerations, and consequences of a particular course of action. To this end, the senior facilities program manager for a department or agency should be directly and continuously involved in the organization's strategic planning processes. This person should be responsible for providing the translation between the agency's mission and its physical assets; identifying alternatives for meeting the mission; identifying the costs, benefits, and potential consequences of the alternatives; and suggesting facilities investments that will reduce overall—that is, portfolio—costs.

RECOMMENDATION 4(a). Each federal department and agency should develop and use a business case analysis for all significant facilities investment proposals to make clear the underlying assumptions, the alternatives considered, the full range of costs and benefits, and potential consequences for the organization and its missions.

There is no standard format for a business case analysis that can be readily adapted directly for use by all federal departments and agencies. However, the committee believes that such an analysis can and should be developed by each federal department and agency and refined over time through repeated, consistent use by the relevant stakeholders and decision makers. At a minimum, a federally adapted business case analysis should explicitly include and clearly state the following: (1) the organization's mission; (2) the basis for the facility requirement; (3)

the objectives to be met by the facility investment and its potential effect on the entire facilities portfolio; (4) performance measures for each objective to indicate how well objectives have been met; (5) identification and analysis of a full range of alternatives to meet the objectives, including the alternative of no action; (6) descriptions of the data, information, and judgments necessary to measure the anticipated performance of the alternatives; (7) a list of the value judgments (i.e., value trade-offs) made to balance achievement on competing objectives; (8) a rationale for the overall evaluation of the alternatives using the information above; (9) strategies for exiting the investment; and (10) the names of the individuals and operating units responsible for the analysis and accountable for the proposed facility's subsequent performance. The business case format to be used by the department or agency should be agreed to by the pertinent oversight constituencies in Congress, the Office of Management and Budget, and the GAO.

RECOMMENDATION 4(b). To promote more effective communication and understanding, each federal department and agency should develop a common terminology agreed upon with its oversight constituencies for use in facilities investment deliberations. In addition, each should train its asset management staff to effectively communicate with groups such as congressional committees having widely different sets of objectives and values. Mirroring this, oversight constituencies should have the capacity and skills to understand the physical aspects of facilities management as practiced in the field.

Engineers, lawyers, accountants, economists, technologists, military personnel, senior executives, and elected officials lack a common vocabulary and style of interaction and do not necessarily share a common set of interests or time frames they consider important. To improve communications among the various stakeholders in facilities investments, each federal department or agency, in collaboration with the appropriate program examiners and congressional representatives, should develop and consistently use a common terminology for the concepts routinely used in facilities investment decision making and applicable to its organizational culture. With the wide variety of missions, cultures, and procedures that exist among federal departments and agencies, a standard set of government-wide definitions is not to be expected.

Training is necessary to ensure that the concepts underlying the terms have meaning and are understood by all. Facilities asset management staff should have the capacity and skills to understand the relationship of facilities to the big picture of an organization's overall mission and to communicate that understanding to others. They should also be able to solve problems by considering all sides of issues and to negotiate a solution that will best meet the organizational requirement. Financial, budget, and program analysts should receive some basic training in facilities investment and management.

RECOMMENDATION 5(a). Each federal department and agency should use life-cycle costing for all significant facilities investment decisions to better inform decision makers about the full costs of a proposed investment. A life-cycle cost analysis should be completed for (1) a full range of facilities investment alternatives; (2) the staff, equipment, and technologies inherent to the alternatives; and (3) the costs of the required funding.

For some very expensive project proposals, federal departments and agencies conduct life-cycle analyses internally to understand the total costs and benefits of the facility itself over the long term and to prioritize their requests for funding. However, in its research and interviews, the committee was not made aware of any instance in which a department or agency also conducted a life-cycle analysis for the staffing, equipment, and technologies inherent to the proposal, or for the life-cycle costs of the required funding.

RECOMMENDATION 5(b). Congress and the administration should jointly lead an effort to revise the budget scorekeeping rules to support facilities investments that are cost-effective in the long term and recognize a full range of costs and benefits, both quantitative and qualitative.

Under federal budget scorekeeping procedures, the budget authority associated with requests to design and construct a new facility, to fund the major renovation of an existing facility, to purchase a facility outright, or to fund operating and capital leases is "scored" up front in the year requested, even though the actual costs may be incurred over several years.

Scoring facilities' costs up front is intended to provide the transparency needed for effective congressional and public oversight. However, implementation of the budget scorekeeping procedures as they relate to facilities investments has resulted in some unintended consequences, including disincentives for cost-effective, long-term decision making and some gamesmanship.

Amending the scorekeeping rules such that they meet congressional oversight objectives for transparency and take into account the long-term interests of departments, agencies, and the public will not be easy. Amending them specifically to account only for life-cycle costs would probably create an even greater disincentive for facilities investments. The committee believes that a collaborative effort that encompasses a wide range of objectives, goals, and values is required. Some possible revisions to the rules could be tested through pilot projects.

RECOMMENDATION 6. Every major facility proposal should include the strategy and costs for exiting the investment as part of its business case analysis. The development and evaluation of exit strategies during the programming process will provide insight into the potential long-

term consequences for the organization, help to identify ways to mitigate the consequences, and help to reduce life-cycle costs.

The development of exit strategies for facilities investment alternatives as part of a business case analysis will help federal decision makers to better understand the potential consequences of the alternative approaches. Evaluation of exit strategies can provide a basis for determining whether it is best to own or lease the required space in a particular situation and whether specialized or more generic "flexible" space is the best solution to meet the requirement. For those investment proposals in which the only exit strategy is demolition and cleanup, evaluating the costs of disposal may lead to better decisions about the design of the facility, its location, and the choice of materials, resulting in lower life-cycle costs.

RECOMMENDATION 7. Each federal department and agency should base its decisions to own or lease facilities on the level of control desired and on the planning horizon for the function, which may not be the same as the life of the facility.

Based on the committee's interviews and research activities, the criteria that departments and agencies use to determine if it is more cost-effective to own or lease facilities to support a given function are not clear or uniform. The committee believes that federal departments and agencies should base the "own" versus "lease" decision on a clearly stated rationale linked to support of the organizational mission, the level of control desired, and the planning horizon for the function to be supported.

RECOMMENDATION 8. Each federal department and agency should use performance measures in conjunction with both periodic and continuous long-term feedback and evaluation of investment decisions to monitor and control investments, measure the outcomes of facilities investment decisions, improve decision-making processes, and enhance organizational accountability.

Because the results of many federal programs or services are qualitative and occur over long periods of time, measuring them can be challenging. However, efforts are under way in various departments and agencies to develop indices and measures that can be applied to evaluate various aspects of facilities portfolios. Some or all of these indices could be adapted for use by other federal departments and agencies and used in combination with other metrics to measure the performance of their facilities' portfolios.

Short-term feedback procedures for facility projects are commonly used. However, to the committee's knowledge, no federal department or agency collected long-term feedback to determine if facilities investments met overall organizational objectives, solved operational problems, or reduced long-term operating costs. Long-term feedback is essential if the outcomes of facilities investments and management processes are to be measured and the decision-making process itself is to be improved.

RECOMMENDATION 9. To increase the transparency of its decision-making process and to enhance accountability, each federal department and agency should develop a decision process diagram that illustrates the many interfaces and points at which decisions about facilities investments are made and the parties responsible for those decisions. Implementation of facilities asset management approaches and consistent use of business case analyses will further enhance organizational accountability.

In the federal government, responsibility and authority for making decisions and executing programs often are not directly linked. Instead, decision-making authority and decision-making responsibility are spread throughout the executive and legislative branches, leading to lack of clear-cut accountability for facilities investment outcomes.

A diagram that illustrates the many interfaces and decision points among the various federal decision-making and operating groups involved in facilities investment decision making can serve as a first step toward increasing the transparency of the process and enhancing accountability. Implementation of a facilities asset management approach, the use of performance measures and feedback processes, and the consistent use of business case analyses will further enhance organizational accountability for federal facilities investments.

RECOMMENDATION 10. Congress and the administration and federal departments and agencies should institute appropriate incentives to reward operating units and individuals who develop and use innovative and cost-effective strategies, procedures, or programs for facilities asset management.

In the federal system, the multiple-objective nature of laws and policies and the sheer volume of procedures sometimes result in unintended consequences, sometimes creating disincentives for good decision making and cost-effective behavior. Potential incentives to support more cost-effective decision making and management by facilities asset management groups could include programs that allow savings from one area of operations to be applied to needs in another area, if the savings are carefully documented; allow the carryover of unobligated funds from one fiscal year to the next for capital improvements, if doing so can be shown to be cost-effective; and establish meaningful awards for operating units with high levels of performance.

RECOMMENDATION 11 (from Chapter 5). In order to leverage funding, Congress and the administration should encourage and allow more widespread use of alternative approaches for acquiring facilities, such as public-private partnerships and capital acquisition funds.

A number of alternative approaches for acquiring facilities are being used by federal departments and agencies, on a case-by-case basis under agency-specific legislation. Each approach has advantages and disadvantages for particular types of organizations and types of facilities. None of the identified alternative approaches can guarantee effective management absent agreed-upon performance measures, feedback procedures, and well-trained staff.

Allowing the use of alternative approaches on a government-wide basis raises concerns about the transparency of funding relationships and concerns about whether the approaches sufficiently account for the perspectives of state and local governments and constituencies. Despite these concerns the committee supports more widespread use of alternative approaches to leverage funding and supports using pilot programs to test the effectiveness of various approaches and to evaluate their outcomes from national, state, and local perspectives. If changes to the budget scorekeeping rules are required to expand the range of alternative approaches, such changes should be tested through the pilot programs.

AN OVERALL STRATEGY FOR IMPLEMENTATION

Transforming decision-making processes, outcomes, and the decision-making environment for federal facilities investments will require sponsorship, leadership, and a commitment of time and resources from many people at all levels of government and from some people outside the government. Implementation of some of the committee's recommendations can begin immediately within federal departments and agencies that invest in and manage significant portfolios of facilities. However, implementing an overall framework of principles and policies will require collaborative, continuing, and concerted efforts among the various legislative and executive branch decision makers and operating groups. These include the President and Congress, senior departmental and agency executives, facilities program managers, operations staff, and budget and management analysts within departments and agencies and from the Congressional Budget Office, the Office of Management and Budget, and the GAO.

Having noted this, the committee is well aware that similar recommendations made by other learned panels advocating long-term, life-cycle stewardship of facilities and infrastructure have achieved only limited success and have failed to move all of the involved stakeholders to action. The committee believes that a new dynamic can and must be instituted and recommends herewith a program it believes practicable.

RECOMMENDED IMPLEMENTATION STRATEGY: The committee recommends that legislation be enacted and Executive Orders be issued that would do two things:

(1) Establish an executive-level commission with representatives from the private sector, academia, and the federal government to determine how the identified principles and policies can be applied in the federal government to improve the outcomes of decision-making and management processes for federal facilities investments within a time certain. The executive-level commission should include representatives from nonfederal organizations acknowledged as leaders in managing large organizations, finance, engineering, facilities asset management, and other appropriate areas. The commission should also include representatives of Congress, federal agencies with large portfolios of facilities, oversight agencies, and others as appropriate. The commission should be tasked to gather relevant information from inside and outside the federal government; hold public hearings; and submit a report to the President and Congress outlining its recommendations for change, an implementation plan, a timetable, and a feedback process for measuring, monitoring, and reporting on the results, all within a time certain.

(2) Concurrently establish department and agency working groups to collaborate with and provide recommendations to the executive-level commission for use in its deliberations. The working groups within each department and agency should collaborate with the executive-level commission. Staff in the departments and agencies are in the best position to communicate their organizational culture and identify practices for implementing the principles and policies that will work for their organization. In addition, they can provide the commission with information related to the characteristics of their facilities portfolios; issues related to aligning their portfolios with their missions; facilities investment trends; good or best practices for facilities investment and management; performance measures for monitoring and measuring the results of investments; and other relevant information.

The committee believes that such sponsorship, leadership, and commitment to this effort will result in

- Improved alignment between federal facilities portfolios and missions, to better support our nation's goals.
- Responsible stewardship of federal facilities and federal funds.
- Substantial savings in facilities investments and life-cycle costs.
- Better use of available resources—people, facilities, and funding.
- Creation of a collaborative environment for federal facilities investment decision making.